A DETAILED HANDBOOK ON FISH PRODUCTION

A DETAILED HANDBOOK ON FISH PRODUCTION

Fish Production Simplified

ANTHONY ADEFARAKAN

Contents

Dedication   1

Acknowledgement   3

Foreword   5

Preface   7

Chapter One   9

Chapter Two   19

Chapter Three   29

                                            43

Chapter Four   59

Chapter Five   77

Chapter Six   83

Chapter Seven   91

Chapter Eight   105

Glossary   111

Reference and Further Reading   115

About the Author   117

Dedication

This book is solely dedicated to my Lord and Savior Jesus Christ, Who alone is the Fountain of all knowledge.

Acknowledgement

I sincerely acknowledge the Almighty God, Who alone is the Source of all wisdom. He is the Author and Finisher of my faith and it is of His fullness that the contents of this book have been drawn.

Also, I appreciate my dear parents, Prince and Mrs. Timothy Olufemi Adefarakan, who sacrificed a great deal to make sound education available to me early in life without showing any signs of fatigue.

My special appreciation also goes to my precious wife, Abisola, whose encouragement and show of understanding provided a conducive environment to get this project accomplished.

Big thanks to my late boss, Ogbueshi Albert Konwea - the Managing Director of Adaeze Farms Ltd, who took his time to drill out the aquaculture expertise resident in me. And also to all my business partners and clients who have engaged my consultancy services at one time or the other.

Foreword

A Detailed Handbook on Fish Production is a book that combines theory with practice. It is an essential guide and instructional guide resource for students, scholars, teachers, researchers and professionals, artisanal fish farmers and those intending to go into fish farming.

The book addresses the basic principles of fish farming, taxonomy, fish production techniques and among others a step-by-step procedure of fish breeding which are rarely treated in conventional texts. It is written in simple language which makes it reader friendly. Also, it is well illustrated and easy to apply.

I found the book so rich that apart from serving as a study guide to students and teachers, practitioners can use it as a working tool in their farms. It also accommodates skills needed by artisanal fish farmers and prospective fish farmers.

I am therefore pleased to recommend this detailed handbook on fish production to high school students, undergraduates, teachers, researchers, professionals including artisanal fish farmers. I see this publication as timely at this period when there is need to improve the productivity of fish production and increase fish output in Nigeria.

Dr. Peter Nwandu (ASN)
Chief Lecturer/Head, Department of Agricultural Education
Federal College of Education (Technical) Asaba,
Delta State, Nigeria.

Preface

Fisheries as it were can be broadly classified into two, namely **capture** and **culture** fisheries. The capture aspect was one of the earliest occupations of man in trying to subdue his environment. This involved setting a trap for fish in any water body without doing anything to improve or replenish the fish stock. It was assumed that the fish stock was inexhaustible, but this has since been proven wrong by the extinction of some fish species.

The development of highly sophisticated capture gears like set nets, trammel nets, spears, line traps, cast nets, and trawlers increased catch per unit effort, thus further depleting the fish stock in the wild. The need to protect young fish stocks and endangered species led to the introduction of control measures like: regulation of mesh sizes, protection of breeding grounds, declaration of closed areas and non-fishing seasons etc. Continuous regulation has helped the situation to some extent; however more still needs to be done.

It was the need to maintain the fish stock while still making fresh fish readily available that led to the development of the culture aspect of fish fisheries otherwise known as aquaculture - which is the rearing of any aquatic flora (plant) and/or fauna (animal) in a controlled environment. It can be done within or outside the fish's natural environment, and it involves the simulation of the natural conditions of various fish species from birth to adulthood. A lot of study had to be done about the physiology, biology, nutrition, reproduction and ideal water condition of various species of fish to achieve this. Aquaculture has gone through a lot of developmental phases; researches have been conducted on a number of aquatic species like catfish, tilapia, salmon, goldfish and shrimps just to mention a few. The production of these species is now being done commercially worldwide.

This book therefore presents a concise and readable interplay between capture and culture fisheries with emphasis on the various processes involved in fish production for both subsistence and commercial purposes. The text, presented with relevant pictures, is intended to provide useful information to the different categories of individuals involved in fish production. Agricultural students in Colleges of Agriculture, Polytechnics, Colleges of Education and Universities, researchers and fish farmers will find this book highly valuable. For detailed treatment of each topic and to ease the readers' understanding, the book is arranged into eight (8) distinct chapters. Each chapter ends with a number of study questions intended to help the reader evaluate the key lessons presented in the chapter. A glossary is also included to aid the understanding of those who are not familiar with fisheries terms as well as references for further reading.

It is believed that this book will not only make the readers become knowledgeable as far as fish production is concerned but also become informed investors in case they wish to engage it as a profitable business venture.

Chapter One

MEANING OF FISHERIES

1. **Fisheries Defined**

   Fisheries simply refer to all processes involved in fish production, processing, marketing and distribution. In other words, the places where fishes are reared for commercial purposes, fishing grounds or areas where fishes are caught as well as the occupation or industry of catching and/or rearing fish are all descriptions of the term *fisheries.*

2. **Sub-divisions of Fisheries**

   Fisheries are broadly classified into two major categories based on the nature of the operations involved. These categories are what constitute the sub-divisions of fisheries; they are *Capture fisheries* and *Culture fisheries*.

   1. **Capture Fisheries**

   Capture fisheries refer to all kinds of harvesting of naturally occurring living resources in both marine and freshwater environments.

   On a broad level, capture fisheries which are sometimes called **wild** fisheries can be classified as industrial, small-scale/ artisanal and recreational.

   The aquatic life they support is not controlled in any meaningful way and needs to be "captured" or fished. Wild fisheries exist primarily in the oceans, and particularly around coasts and continental shelves. They also exist in lakes and rivers.

Fig. 1.1: Crab boat from the North Frisian
Islands working in the North Sea

Certain issues are however associated with wild fisheries. In the process of fishing, the following effects usually result;

1. **Habitat destruction**
   In the course of fishing, certain fishing nets are left or lost in the ocean by fishermen, these are called ghost nets and they can entangle fish, dolphins, sea turtles, sharks, dugongs, crocodiles, seabirds, crabs, and other creatures. Acting as designed, these nets restrict movement, causing starvation, laceration and infection, and—in those that need to return to the surface to breathe—suffocation.

2. **Overfishing**

This refers to a situation where fishing is carried out without control, especially in a particular water body. Some specific examples of overfishing are described below;

- On the east coast of the United States, the availability of bay scallops has been greatly diminished by the overfishing of sharks in the area. A variety of sharks have, until recently, fed on rays, which are a main predator of bay scallops. With the shark population reduced, in some places almost totally, the rays have been free to dine on scallops to the point of greatly decreasing their numbers.

- Chesapeake Bay's once-flourishing oyster populations historically filtered the estuary's entire water volume of excess nutrients every three or four days. Today that process takes almost a year, and sediment, nutrients, and algae can cause problems in local waters. Oysters filter these pollutants, and either eat them or shape them into small packets that are deposited on the bottom where they are harmless.

  ◦ The Australian government alleged in 2006 that Japan illegally overfished southern bluefin tuna by taking 12,000 to 20,000 tonnes per year instead of their agreed 6,000 tonnes; the value of such overfishing would be as much as US$2 billion. Such overfishing has resulted in severe damage to stocks. "Japan's huge appetite for tuna will take the most sought-after stocks to the brink of commercial extinction unless fisheries agree on more rigid quotas" stated the World Wide Fund for Nature (WWF). Japan disputes this figure, but acknowledges that some overfishing has occurred in the past. These are just to mention a few. Significant wild fisheries have collapsed or are in danger of collapsing, due to overfishing and pollution. Overall, production from the world's wild fisheries has levelled out, and may be starting to decline.

1. **Loss of biodiversity**

Each species in an ecosystem is affected by the other species in that ecosystem. There are very few single prey-single predator relationships. Most prey are consumed by more than one predator, and most predators have more than one prey. Their relationships are also influenced by other environmental factors. In most cases, if one species is removed from an ecosystem, other species will most likely be affected, up to the point of extinction.

Species biodiversity is a major contributor to the stability of ecosystems. When an organism exploits a wide range of resources, a decrease in biodiversity is less likely to have an impact. However, for an organism which exploits only limited resources, a decrease in biodiversity is more likely to have a strong effect.

Reduction of habitat, hunting and fishing of some species to extinction or near extinction, and pollution tend to tip the balance of biodiversity.

2. **Threatened species**

The global standard for recording threatened marine species is the International Union for Conservation of Nature (IUCN) Red List of Threatened Species. This list is the foundation for marine conservation priorities worldwide. A species is listed in the threatened category if it is considered to be critically endangered, endangered, or vulnerable. Other categories are near threatened and data deficient. This will be considered in both marine and freshwater habitats.

• **Marine**

Many marine species are under increasing risk of extinction and marine biodiversity is undergoing potentially irreversible loss due to threats such as overfishing, bycatch, climate change, invasive species and coastal development.

By 2008, the IUCN had assessed about 3,000 marine species. This includes assessments of known species of shark, ray, chimaera, reef-building coral, grouper, marine turtle, seabird, and marine mammal. Almost one-quarter (22%) of these groups have been listed as threatened. Table 1.1 below gives the details.

Group	Species	Threatened	Near threatened	Data deficient
Sharks, rays, and chimaeras		17%	13%	47%
Groupers		12%	14%	30%
Reef-building corals	845	27%	20%	17%
Marine mammals		25%		
Seabirds		27%		
Marine turtles	7	86%		

Table 1.1: Data Analysis of certain marine species assessed by IUCN in 2008
(Source: Wikipedia)

- **Sharks, rays, and chimaeras**: are deep water pelagic species, which makes them difficult to study in the wild. Not a lot is known about their ecology and population status. Much of what is currently known is from their capture in nets from both targeted and accidental catch. Many of these slow growing species are not recovering from overfishing by shark fisheries around the world.
- **Groupers**: Major threats are overfishing, particularly the uncontrolled fishing of small juveniles and spawning adults.
- **Coral reefs**: The primary threats to corals are bleaching and

disease which has been linked to an increase in sea temperatures. Other threats include coastal development, coral extraction, sedimentation and pollution. The coral triangle (Indo-Malay-Philippine archipelago) region has the highest number of reef-building coral species in threatened category as well as the highest coral species diversity. The loss of coral reef ecosystems will have devastating effects on many marine species, as well as on people that depend on reef resources for their livelihoods.

- **Marine mammals**: include whales, dolphins, porpoises, seals, sea lions, walruses, sea otter, marine otter, manatees, dugong and the polar bear. Major threats include entanglement in ghost nets, targeted harvesting, noise pollution from military and seismic sonar, and boat strikes. Other threats are water pollution, habitat loss from coastal development, loss of food sources due to the collapse of fisheries, and climate change.
- **Seabirds**: Major threats include <u>longline fisheries</u> and gillnets, oil spills, and predation by rodents and cats in their breeding grounds. Other threats are habitat loss and degradation from coastal development, logging and pollution.
- **Marine turtles**: Marine turtles lay their eggs on beaches, and are subject to threats such as coastal development, sand mining, and predators, including humans who collect their eggs for food in many parts of the world. At sea, marine turtles can be targeted by small scale subsistence fisheries, or become <u>bycatch</u> during <u>longline</u> and trawling activities, or become entangled in ghost nets or struck by boats.

- **Freshwater**

Freshwater fisheries have a disproportionately high diversity of species compared to other ecosystems. Although freshwater habitats cover less than 1% of the world's surface, they provide a home for over 25% of known vertebrates, more than 126,000 known animal species, about 24,800 species of freshwater fish, molluscs, crabs and dragonflies, and about 2,600 macrophytes. Continuing industrial and agricultural developments place huge strain on these freshwater systems. Waters are polluted or extracted at high levels, wetlands are drained, rivers channelled, forests deforestated leading to sedimentation, invasive species are introduced, and over-harvesting occurs.

In the 2008 IUCN Red List, about 6,000 or 22% of the known freshwater species have been assessed at a global scale, leaving about 21,000 species still to be assessed. This makes clear that, worldwide, freshwater species are highly threatened, possibly more so than species in marine fisheries. However, a significant proportion of freshwater species are listed as data deficient, requiring more field surveys.

Capture fisheries, despite serving as means for sustenance of livelihood for subsistence fishermen, sources of revenue for industrial fishermen and means of relaxation through recreational fishing, have devastating effects on the aquatic organisms as well as the habitat as highlighted above.

1. Culture Fisheries

Culture fisheries, otherwise referred to as farmed fisheries or aquaculture, involve the rearing or production of aquatic organisms in a controlled water environment.

This implies an act of fish husbandry whereby fish are kept and fed in confinement and managed to achieve quick growth and accelerated reproduction.

As a contrast to wild fisheries, farmed fisheries can operate in sheltered coastal waters, in rivers, lakes and ponds, or in enclosed bodies of water such as tanks. Farmed fisheries are technological in nature, and revolve around developments in aquaculture. Farmed fisheries are expanding, and Chinese aquaculture in particular is making many advances.

Aquaculture has continued to thrive globally over the years mainly due to its simplified technology of operation and availability of artificially bred fish seeds.

Fig. 1.2 below gives a graphical representation of the advances in aquaculture as against

capture fisheries up to the early part of the 21st century.

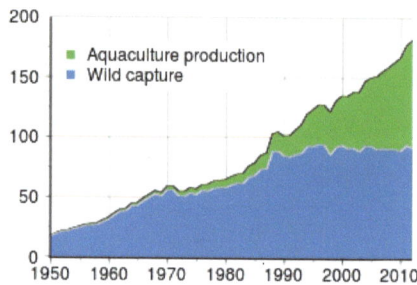

Fig. 1.2: Global harvest of aquatic organisms in million tonnes, 1950–2010, as reported by the FAO
(Source: Based on data from the Fish Stat database)

Certain factors however come to play in determining fishes that can be cultured as not all species are culturable.

Such culturable fish species must therefore possess the following characteristics;

- **Hardiness** – the fish must be hardy. It must be able to withstand handling conditions and be adaptive to the associated stress.
- **High Survival Rate** – a large number or percentage of the fish population must be able to survive in captivity.
- **Compatibility** – the fish must be able to co-exist with other fish species. It must be adaptive to polyculture practice.
- **Acceptance of Artificial Feed** – the species must be able to accept formulated or artificially compounded feed.

- **Breeding under Captivity** – the fish must be able to reproduce artificially. It must be able to breed in confinement, that is, in an artificial environment.
- **Tolerance to Poor Water** – culturable fish species must be able to grow and thrive under conditions of poor water quality.
- **Fast Growth** – the fish species must have good growth rate and reach market size within a short period of time. That is, it must have high feed conversion efficiency.
- **Disease Resistance** – such species must be able to withstand a wide range of fish diseases.
- **Acceptability to Consumers** – the fish must be marketable. It must be appealing and acceptable to the target consumers.
- **Socio –cultural Acceptability** – such species must not constitute any socio-cultural nuisance. It must not be against the religious or traditional beliefs of the community or environment where it is to be cultured. It must not attract any religious, social or cultural bias.

These are basic factors to consider before listing any fish species as culturable. Other important facts about culture fisheries such as culture facilities, aquaculture management operations among others are treated in detail in Chapter Three.

Study Questions:

1. Define the term *fisheries*.
2. State four (4) demerits associated with capture fisheries.
3. Certain characteristics make some fish species culturable, mention any six (6) of such characteristics.

Chapter Two

FISH TAXONOMY

2.1 Taxonomical Classification of Fish

Taxonomy is the practice and science of classification. It is composed of two Greek words, *taxis* (order, arrangement) and *nomos* (law or science); hence the word literally means "Science of arrangement".

The first attempts to classify fish began in ancient world, the first recorded work is in Indian text known as "Kautila's Arthasastra" belonging to period about 321 BC.

These were very detailed descriptions, like "Pirthuroman" which means a fish with long hairs, i.e. barbels near snout. Around 246 BC King Asoka had pillars erected throughout India, these pillars had vital information inscribed on them. Many of them had descriptions of various animals including fish. Some examples are:

1. Anathikamacchi- boneless fish (sharks)
2. Vedaveyake- eluding grasp (Eels)
3. Kaphatasayake- sleep feigning fish (puffer fish).

Hippocrates and Aristotle from around 384-322 BC from Greece are two other pioneers who developed animal classification.

There have been changes in the classification system; however the system proposed by Carl Linne (Linnaeus) is followed till today.

He proposed what is called bi nominal nomenclature, naming every living thing with a two part name, the first one based on genus(plural genera) and second part on species (same in singular or plural).

This system again saw a major change with Charles Darwin's publication "The origin of Life" in 1859.

His theory of evolution meant that species placed together in a genus were assumed to have had a common origin, a concept that underlies all important subsequent classifications of fishes and other organisms in the current system.

The current system is based on classifying living beings into separate divisions called "Taxon" or "Taxa" in plural, hence the term **Taxonomy.** Each level of this classification is called a "Rank".

The classification used today must conform to "International Code of Zoological Nomenclature" which is administered by International Commission on Zoological Nomenclature based at "Natural History Museum" in London.

The highest rank is called "Kingdom". There are five kingdoms currently recognized, Monera, Protista, Plantae, Fungi, and Animalia. All animals are grouped under kingdom "Animalia".

Fish are members of the Animalia Kingdom (animals) and are also classified into the Phylum Chordata. In order to be a chordate an animal must have a notochord (a slim and flexible rod that supports the body) at some point in their lives; have a tubular nerve chord along their back (dorsal surface) with the brain developing from a swelling found at the anterior end (front) of this tube; paired gill slits at some stage of their life history; segmentation of at least part of their body; a post-anal tail at some stage in their life history; a ventral heart; and an endoskeleton.

Fish are further classified into the Vertebrata Subphylum. In order to be a vertebrate, an animal must have a vertebral column, or backbone. This backbone encloses, supports and protects the spinal cord. Fish are vertebrates that live in water and breathe with gills. Fish are ectotherms, or cold-blooded. Fish have either backbones of cartilage or bone. Most fish are adapted to live in salt or fresh water. Most fish have fins and scales, which cover and protect the body. The body systems of the fish, such as the digestive, circulatory are well developed. The fish are further classified into three classes. These classes are the Agnatha, jawless fish such as the hagfish and lampreys; the Chrondrichthyes, fish whose skeleton is made of cartilage such as sharks, rays and skates; and the Osteichthyes, fish whose skeleton is composed mostly of bone such as bass, perch, catfish, and flounder. Below are some examples of their subclasses;

- Class Agnatha (jawless fish)
    ◦ Subclass Cyclostomata (hagfish and lampreys)
    ◦ Subclass Ostracodermi (armoured jawless fish)

Fig.2.1.1: Agnatha (Pacific hagfish)

- Class Chondrichthyes (cartilaginous fish)
    ◦ Subclass Elasmobranchii (sharks and rays)
    ◦ Subclass Holocephali (chimaeras and extinct relatives)

Fig.2.1.2: Chondrichthyes (Horn shark)

- Class Osteichthyes (bony fish)
  - Subclass Actinopterygii (ray finned fishes)
  - Subclass Sarcopterygii (fleshy finned fishes, ancestors of tetrapods)

Fig.2.1.3: Sarcopterygii (Coelacanth)

Fig.2.1.4: Actinopterygii (Brown trout)

There are approximately 50 species of jawless fish, 600 species of car-tilaginous fish and more than 30,000 species of bony fish. The bony fish,

Osteichthyes, are then further classified into two main groups called the ray-finned group (perch, and catfish) and the lobe-finned group (lungfish). Most bony fish belong to the "ray-finned" group. There are approximately 70 fish orders known to biologists. Further examples on fish taxonomical classifications are given below;

2.1.1 Fish Classes
Kingdom Animalia --animals
Phylum Chordata -- chordates
Subphylum Vertebrata --vertebrates
Superclass Osteichthyes -- bony fishes
Class Actinopterygii --ray-finned fishes, spiny rayed fish
Subclass Neopterygii -- neopterygians
Infraclass Teleostei
Superorder Acanthopterygii
Order Perciformes -- perch-like fishes
Suborder Percoidei
Family Centrarchidae --sunfishes
Genus Micropterus Lacepède -- black basses, largemouth basses
Species *Micropterus salmoides*

2.1.2 Fish Orders
Order Lepisosteiformes
Order Acipenseriformes (Sturgeons and Paddlefishes)
Order Cypriniformes (Minnows, Carps and Suckers)
Order Siluriformes (North American Catfishes)
Order Esociformes (Pikes and Pickerels)
Order Salmoniformes (Trout and Salmon)
Order Scorpaeniformes (Sculpins)
Order Perciformes (Temperate Basses, Sunfish, Perches, Drums)

2.2 Fish Breeds
There are several thousands of fish species in existence, varying in colors, sizes, feeding patterns, habitat conditions, body

structures among other distinguishing features. This in essence points to the fact that there is great diversity of fish.

A typical fish is ectothermic, has a streamlined body for rapid swimming, extracts oxygen from water using gills or uses an accessory breathing organ to breathe atmospheric oxygen, has two sets of paired fins, usually one or two (rarely three) dorsal fins, an anal fin, and a tail fin, has jaws, has skin that is usually covered with scales, and lays eggs.

Fish range in size from the huge 16-metre (52 ft) whale shark to the tiny 8-millimetre (0.3 in) stout infant fish.

Fish species diversity is roughly divided equally between marine (oceanic) and freshwater ecosystems. Coral reefs in the Indo-Pacific constitute the center of diversity for marine fishes, whereas continental freshwater fishes are most diverse in large river basins of tropical rainforests, especially the Amazon, Congo, and Mekong basins. More than 5,600 fish species inhabit Neotropical freshwaters alone, such that Neotropical fishes represent about 10% of all vertebrate species on the Earth. Exceptionally rich sites in the Amazon basin, such as Cantão State Park, can contain more freshwater fish species than occur in all of Europe.

For ease of study however, fish breeds have been classified into five broad categories based on certain characteristics. These are further described below.

2.2.1 Classification Based on Habitats

1. **Freshwater Fish** – these are fish that spend some or all of their lives in fresh water such as rivers and lakes with a salinity of less than 0.05%. Examples of such species are *Clarias gariepinus*, *Heterotis niloticus*, *Tilapia zilli* and other ornamental species like the ones represented in the pictures below:

Fig. 2.2.1: Barbs

Fig. 2.2.1: Convict
Cichlid (Zebra fish)

Fig. 2.2.3: Goldfish

2. **Saltwater Fish** –these fish spend some or all of their lives in salt water such as oceans and salt lakes, generally with a salinity of more than 0.05%. Examples of such breeds are Mackerel, Coley, Cod, and some ornamental (aquarium) breeds such as those in the pictures below:

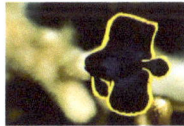

Fig. 2.2.4: Batfish
(Saltwater)

Fig. 2.2.5: Snappers

3. **Brackish water Fish** – these fish breeds live in areas where freshwater meets saltwater. Such fish are able to tolerate a wide range of salinities up to 1.015% and a pH of 7.5 – 8.4. Examples include *Sarotherodon melanotheron, Eutroplus maculatus* etc

2.2.2 Classification Based on Anatomy

a. Bony Fish – these are fish with bony skeletal framework, cycloid scales and apical mouth part. Examples include *Tilapia spp, Heterobranchus spp, Oreochromis niloticus, Cyprinus carpio, Lates niloticus, Oreochromis aureus, Heterotis niloticus* etc

b. Cartilaginous Fish – these are fish with paired fins, placoid scales, two chambered hearts, ventral mouth and skeleton made up of cartilage. Examples include Sharks and Rays.

2.2.3 Classification Based on Feeding Patterns

a. Herbivorous Fish – these are fish which feed on plants, especially aquatic weeds such as emergent macrophytes, submerged macrophytes, floating macrophytes and algae. Examples of such species are *Ctenopharyngodon idella* (Grass carp) and *Oreochromis niloticus* (Nile tilapia).

b. Carnivorous Fish – these feed on other animals in the water body especially zooplankton and other fish species. Examples include *Clarias gariepinus, Heterobranchus bidorsalis, Heterobranchus longifilis* etc

c. Omnivorous Fish – these feed on both plants and animals. Examples include *Cyprinus carpio* (Common carp), *Cirrhinus molitorella* (Mud carp), Sharks, Whales and Salt water lobsters.

2.2.4 Classification Based on Environment Utilized in the Water Body

a. Pelagic Fish – these fish mostly utilize the surface part of the water; they are usually referred to as surface dwellers. Examples include *Tilapia zilli, Oreochromis niloticus, Oreochromis aureus* etc

b. Benthic Fish – these usually utilize the bottom part of the water; otherwise known as bottom dwellers. Examples include *Clarias gariepinus, Heterobranchus spp, Cyprinus carpio, Mylopharyngodon piceus* (Black carp) etc

c. Benthopelagic fish- these utilize the part of the water just above the bottom. Examples of these include the deep sea cods (morids), deep sea eels, deep sea squaloid sharks etc

2.2.5 Classification Based on Climatic Regions

a. Temperate Fish – these are fish species cultured or raised in climates with mild or cold temperature. They are well adapted to this cold condition. Examples include Salmon and Mackerels.

b. Tropical Fish – these are fish that can be cultured in regions with considerably hot climate. They thrive under such thermal condition. Examples include *Cyprinus carpio, Tilapia zili, Clarias gariepinus* etc.

Study Questions:

1. Briefly describe the term *taxonomy*.
2. Give two (2) examples each of fishes of the Class Agnatha and Class Chondrichthyes respectively.
3. Differentiate between *pelagic* and *benthic* fishes.

Chapter Three

PRINCIPLES OF FISH FARMING

3.0 Introduction to Aquaculture

Aquaculture is the rearing of any aquatic flora (plant) and/ or fauna (animal) in a controlled environment. Fish farming can be broadly classified as **culture fisheries** (as stated in chapter one). The capture aspect of fisheries was one of the earliest occupations of man in trying to subdue his environment. This involved getting fish out of water bodies without doing anything to improve or replenish the stock. It was assumed that the fish stock was inexhaustible, but this has since been proven otherwise by the extinction of some fish species.

The development of highly sophisticated capture gears like set nets, trammel nets, spears, line traps, cast nets, and trawlers increased catch per unit effort, thus further depleting the fish stock in the wild. The need to protect young fish stocks and endangered species led to the introduction of control measures like: regulation of mesh sizes, protection of breeding grounds, declaration of closed areas and non-fishing seasons etc. Continuous regulation had helped the situation to some extent; however certain stocks are still endangered due to this method of fisheries.

It was the need to maintain the fish stock in the wild and yet making fresh fish readily available that led to the development of the culture aspect of fisheries otherwise known as **AQUA-CULTURE** (which has been earlier defined). It can be practised within or outside the fish's natural environment by simulating the natural condition of various culturable fish species from birth to adulthood. A lot of research has been carried out on the physiology, biology, nutrition, reproduction and ideal water condition of various species of fish for this to be achieved. Aquaculture has gone through a lot of developmental phases; researches have been conducted on a number of aquatic species like catfish, tilapia, salmon, goldfish, shrimps just to mention a few. As a matter of fact, the first attempt at fish farming in Nigeria was in 1951 at a small experimental station in Onikan, Lagos State culturing *Tilapia* species. Eventually, pilot farms were established in Panyam, Plateau State for rearing the Common (mirror) carp, *Cyprinus carpio* which consequently generated sufficient interest to encourage regional governments to establish more fish farms. The production of these species is now being done commercially not only in Nigeria but also globally.

3.1 General Operating System (Fish Culture Systems)

This is classified into three on the basis of the degree of control. The three operating systems are extensive, semi-intensive and intensive systems.

1. **Extensive System:** In this system, only fish seeds i.e. fingerlings or juveniles are introduced into the culture medium; water quality management, nutrition, size sorting and other means of control are not done. Many fish species are usually stocked together (polyculture) so that the bigger ones can prey on the younger (smaller) ones.

   Extensive systems are also exposed to sunlight, thus enhancing photosynthesis by phytoplanktons. This consequently helps the bloom of zooplanktons like daphnia, moina etc. The phyto-planktons and zooplanktons also serve as food to the stocked fish. Extensive aquaculture is usually practiced in dams, lakes or earthen ponds. It is however usually characterized by slow and uneven growth, low survival rate due to poor management and predation, thus giving very low fish yield.

2. **Semi-Intensive System:** This is what is widely practiced in Nigeria today. It is steps away from the extensive system where little or no control is done. Here, virtually all aspects of the fish culture is controlled from the stocking rate to nutritional con-trol to water quality management to predation control by con-sistent size grading which are done when due. This system can be practiced in earthen ponds, concrete tanks, plastic tanks, glass tanks, wooden vats etc. The only snag in the system is that aeration and waste management are not efficiently done, water is allowed to flow through the fish ponds at intervals to give supplementary oxygen while removing some of the liquid and solid wastes. But when effectively practiced, semi-inten-sive systems usually give good yield to justify the input.

3. **Intensive System :**As the name suggests, the degree of control here is high. Virtually all aspects of production here are reg-ulated. Fish stocking rate is high, feeding is adequately done, aeration is consistently done, waste removal is efficiently done, fish size regulation is consistently done. This system is usually called the water recirculatory system, where attempts are made to conserve the water use by allowing the water to go through bacteria culturing media in which ammonia is converted to ni-trite and then to nitrate.

The various components of an intensive system are:

1. **Fish Holding Tanks**: these are usually made of concrete tanks, fibre glass tanks, wooden vats etc. They should be strong

enough to hold the high fish density. The fish tank must have an efficient water inlet and outlet cum waste removal mechanism.

2. **Clarifying Chamber:** water carrying solid waste from the fish tank goes through a solid filtration bed called the sedimentation or clarifying chamber. This is made of various grades and types of sieves installed in layers so that different sizes of solids are removed as the water goes through. The sedimentation chamber must be back flushed regularly to avoid ammonia build up there. The frequency of flushing will depend on the size of the chamber and the fish stocking density in the system.

3. **Water Pumping Tank:** this tank is usually necessary where the tower form of bio-filter is in use. Water from the clarifying chamber overflows into the pump tank, it is then taken to the top of the bio-tower by an electrical pumping machine. The pump tank also serves as the dilution chamber as fresh water is introduced into the system from here.

4. **The Bio-filter:** this is the back bone of intensive aquaculture. Nitrosomonas and Nitrobacter bacteria are cultured here. These convert ammonia to nitrite and nitrite to nitrate. Ammonia concentration should not be more than 3mg per litre, nitrite concentration not higher than 2mg per litre at any point in time. High concentrations of nitrate can however be tolerated by catfish. The biofiltration chamber is usually raised above the fish tank so that oxygen can be added to the water while falling through the tower and water can go back to the fish tank by gravity.

The intensive aquaculture system adopts a series pond connection thus ensuring similar conditions in all the connected ponds. The disadvantage of this however could be rapid spread of infection into all the tanks in case of an outbreak. The system could also be quite expensive as a result of its dependence on electricity to run the pumps. The water pumping machine should be used throughout the day to prevent ammonia and nitrite build up during periods of stagnation.

Generally, the fish culture system is also determined by the species and sex of fish stocked. For example, if a single species of fish is cultured, this system is known or described as monoculture system while more than one species that are raised simultaneously in the same holding facility (ponds or fish tanks) is described as polyculture. Polyculture system operates or capitalizes on using different food niches within the environment to raise complementary fish species. Up to six species of fish have been developed in a model called China polyculture system. If a single sex of a particular species is cultured, it is known as monosex. Culture system could as well be described by the water holding facility and the water exchange method. In the first case, we could have earthen pond culture system, cage culture system, concrete tank culture system etc while in the second case we could have flow through system, water recycling pond/tank culture system or static water culture system. Having these varieties, a fish farmer will therefore need to consider and choose the operating or culture system that best suits his production plan.

Below are pictures of some culture facilities a farmer may wish to engage in practicing aquaculture;

Fig. 3.1.1a: Collapsible Tarpaulin Fish Tank

Fig. 3.1.1b: Mobile Tarpaulin Fish Tank

Fig.3.1.2a: Outdoor Plastic Fish Tanks

Fig.3.1.2b: Indoor Plastic Fish Tanks

Fig.3.1.3a: Concrete Fish Tank

Fig.3.1.3b: Concrete Fish Tank

Fig.3.1.4a: Earthen Fish Ponds

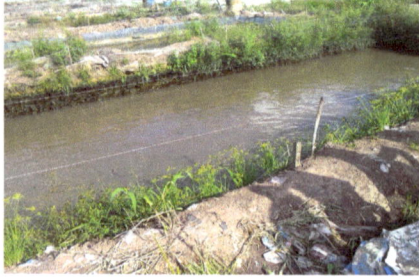

Fig.3.1.4b: Earthen Fish Ponds

3.2 Fish Farm Establishment

Fish farming requires a thorough analysis of the feasibility of the project before its commencement. Aquaculture projects are many and vary with regards to the type to be practiced (as earlier stated). The feasibility report should demonstrate that the aquaculture project will be successful. This is because even if the fish farm project is well located, fish ponds well-constructed and it is run by experts or fish professionals, without a well thought-out feasibility study, the project is bound to fail.

For any aquaculture project, there are three broad areas of feasibility to be analyzed and documented for the success of the project. (i)the location feasibility (ii) the biotechnical feasibility and (iii) the economic feasibility.

3.2.1 Feasibility Studies in Aquaculture

Feasibility, as earlier stated, is a measure of a project's or business' viability or success. In aquaculture, feasibility of a fish farm can exclude locational feasibility if the project does not involve an earthen pond or if it is going to be executed in an urban area. However, the bio-technical and economic feasibility studies are very important to the success of the project.

3.2.1.1 Location Feasibility

This is the analysis of various factors that are important in choosing a good location for the establishment of a viable fish farm. It involves the study of the topography (vegetation, rocks, slopes, contours etc), water source (its quality and quantity/ availability), and other physical characteristics of the environment. In citing an aquaculture project, it is very important to know the characteristics of the land (soil types, direction of the slope, vegetation of the area and how high or low the land is). Therefore, for a fish culture project, the contour map of the location is required, showing the full drainage pattern and land levels. This contour map will make it easy for the experts to know how to design the fish farm. The structure of the soil and its composition need to be studied carefully to be able to know the water retention capacity of the soil. The composition and structure of soils vary from place to place, and guess work on whether it's sandy, loamy or clay soil does not work in aquaculture. For instance, sandy soil, gravel soil and soils that cannot hold water all year round are not suitable for fish ponds. Also, the study of the composition of the soil shows whether it is acidic or alkaline and the level of acidity and alkalinity.

In an intensive or semi-intensive aquaculture project, especially in the urban areas, characteristics like vegetation, soil types, rocks, slopes, contours etc are not of much importance. However, water source (its quantity and quality) is very important to locating fish ponds in urban areas.

Fish can only live and grow in water, therefore, if water is not available in the proposed area there cannot be fish farming no matter how small the project is. Hence, the source of water to be used, its quality and quantity, is very important and must be studied carefully before a fish farm is established. Not all water is good for aquaculture, hence the source of the water (well, borehole, streams, rivers etc); its quality (alkalinity or acidity, dissolved oxygen properties etc); how far or near it is from site of proposed fish farm; the means of procurement to the farm site (pumping with machine, natural gravity flow through slopes), the cost and technology of procurement of water, presence or absence of pollutants or whether purification is required, the nature of industries nearby with regards to their waste discharge (whether polluting or could be harmful to the fish) among others must be critically examined and thoroughly analyzed.

3.2.1.2 Biotechnical Feasibility

This involves the study of factors that determine the type of fish farm to be established. It involves selection of species, pond design and engineering and the kind of operating system to be employed. Here, a number of questions are expected to be answered as regards the desired species to be cultured on the fish farm, the best and affordable design of fish pond with its mode of construction and the kind of practice (intensive, semi intensive or extensive).

1. **Species Selection**

    The desired species to be cultured on the farm must be studied with regards to its availability locally, its suitability for such local culture, its ability to withstand environmental stress, its food and feeding habits, growth rate, reproduction pattern, resistance to disease attack and availability of its fingerlings. If a wrong species is chosen, it will lead to the failure of the fish farm. Certain features characterize culturable fish species as highlighted in chapter one; these are to be carefully considered in the choice of the species to culture.

2. **Pond Design and Engineering**

    The kind of pond for fish culture in any fish farm project depends on the scale of the project, whether small scale, medium scale or large scale. Basically, three designs of fish ponds are common in aquaculture namely: the earthen ponds (channel ponds, barrage ponds etc), concrete ponds (flow through) and recirculatory ponds. Their designs and engineering works depend on the choice of the fish farmer. The farmer has the choice of the type of design to use for the fish culture. Usually earthen ponds are used in the rural areas where there is expansive land for such practice, as well as good streams and rivers as water sources. In urban areas, the use of concrete and mobile tanks is common, with the trending use of the recirculatory systems for large scale aquaculture practice. Boreholes and wells are regular sources of water for such practices in urban areas.

3.2.1.3 Economic Feasibility

This is the most important aspect of feasibility study for fish farm establishment, especially when it is designed for commercial purpose. If this aspect is faulty or absent, the whole project is bound to fail, even if the fish farm is well located with best facilities, excellent designs and perfect construction coupled with good technical management practices.

Economic feasibility deals with analyses of financial inputs, output projections and market survey and analyses. It is in the course of this analysis that the farmer decides whether the project is viable and worth venturing into or to simply withdraw. It must demonstrate that the inputs will give optimal output and good profit margin, considering the market analyses done. If the feasibility cannot show such result at the end of one year (or maximum of two years), then it is not worth it to proceed with such fish farm project.

In assessing financial inputs that go into any fish farm, every item on which expenditure is made must be critically analyzed. Such inputs, usually divided into two types; fixed inputs and variable inputs, are the most important components of fish production process. The fixed inputs are those that are constant during fish farm establishment, usually depreciated over a period of five to ten years (depending on the choice of the fish farmer). Fixed inputs include land, fish tanks/ponds, farm house, borehole or well, generators etc.

Variable inputs are those that change with season, level of production and management practices, they are used in execution of the project. These include fish fingerlings, fish feeds, fertilizers, labour, electricity bills, fuel, water, tax etc.

In the assessment of the economics of establishing a fish farm, the expenditure for each item must be given possible options that can still yield maximal output. For instance, the option of either producing fish fingerlings on the farm is weighed against purchasing fingerlings from other well established fish farms. Similarly, fish feed may be assessed based on its purchase as imported feeds, locally compounded feeds (based on specifications) or feeds produced on the farm. Thus, the cost, advantages and disadvantages of each of these options must be critically analyzed before a decision is made.

An economic feasibility must be able to make reasonable and realistic projections on the possible income from the output of a fish farm project. It should include the critical analysis of existing fish market, with regards to the customers, pricing, value attached to each species of fish, biases, preferences etc. It should include a credible and reasonable cash flow projection, based on existing knowledge of the market and a projection of future trends. The existing sources of fish supply, the quantity, quality, seasonality and species available in the market must be assessed as well.

This feasibility also takes into consideration the fish demand pattern, its flexibility (or rigidity) and its effect on future expansion of the fish farm project. Furthermore, possible alternatives to the local market should be determined by the feasibility, and it should indicate the necessity of advertisement in expanding the market or creating new markets, dissolving existing biases etc. '

The following are examples of the cost implication of establishing a 6 tank recirculatory system fish farm, with income projections for first year of operation.

Note: These may not be the recent prices of the items.

Table 3.1 Capital Expenditure (Fixed Input)

S/No	Item	Qty	Rate (N : K)	Amount (N : K)
1	9" x 9" x 18" vibrated blocks	500	150.00	75,000.00
2	6" x 9" x 18" vibrated blocks	1,000	130.00	130,000.00
3	Sharp sand	6 tippers	12,000.00	72,000.00
4	Soft sand	6 tippers	14,000.00	84,000.00
5	Gravel	6 tippers	24,000.00	144,000.00
6	Cements	180 bags	2,600.00	468,000.00
7	Biofilter	1	250,000.00	250,000.00
8	Plumbing materials/ labour			250,000.00
9	Labour: Bricklayers			200,000.00
10	Wiring/ Electrical expenses			50,000.00
11	Generator (5KVA)			220,000.00
12	Borehole/ deep well			350,000.00

13	Project supervision by professionals			120,000.00
	Sub Total			**2,413,000.00**

S/No	Item	Amount (N : K)
1	Fertilization and water analysis	20,000.00
2	Dutch catfish fingerlings @ N15 x 25,000 pieces	375,000.00
3	Feeding and vaccines (N180 x 1kg x 25,000)	4,500,000.00
4	Electricity bills (N5,000 x 6 months)	30,000.00
5	Fuel @ N48.50 x 2,000 litres	97,000.00
6	Salaries/wages (N12,500 x 2 x 6 months)	150,000.00
7	Consultancy (N30,000 x 6 months)	180,000.00
8	Contingencies	77,000.00
	Sub Total	5,449,000.00
	TOTAL	N6,950,000.00

INCOME STATEMENT

Assumed survival rate – 90%

Sales: 22,500 x 1kg x N350 = N7,875,000.00

Less capital cost depreciation (10%): N 152,100.00

=N7,722,900.00

Less recurrent expenditure: N5,449,000.00

Net profit = N2,273,900.00

Table 3.3 Recurrent Expenditure (Variable Input) - (SECOND PRODUCTION)

S/No	Item	Amount (N : K)
1	Fertilization and water analysis	20,000.00
2	Dutch catfish fingerlings @ N15 x 25,000 pieces	375,000.00
3	Feeding and vaccines (N180 x 1kg x 25,000)	4,500,000.00
4	Electricity bills (N5,000 x 6 months)	30,000.00
5	Fuel @ N48.50 x 2,000 litres	97,000.00
6	Salaries/wages (N12,500 x 2 x 6 months)	150,000.00
7	Consultancy (N30,000 x 6 months)	180,000.00
8	Contingencies	77,000.00
	Sub Total	**5,449,000.00**

INCOME STATEMENT

Assumed survival rate – 95%

Sales: 23,750 x 1kg x N350 =N8,312,500.00

    Less capital cost depreciation (10%): N 152,100.00

= N8,160,400.00

    Less recurrent expenditure :N5,449,000.00

    Net profit = N2,711,400.00

A farmer should always check the profit margin ratio when making financial or economic projections of the project. This is given by: **Net Profit/Turn over X 100.** It is expressed in percentage.

The Profit margin is good when a high percentage of the turnover is converted to net profit usually by reducing production costs by all means while at the same time increasing sales. The higher the percentage, the more profitable the business promises to be.

3.3Fish Farm Management and Operations

This is of much importance as poor management structure and practices can frustrate the whole system no matter how good the project is. In any fish farm project that is to be successful, the management structure should give the hierarchy of staff, their duties clearly defined and their remuneration paid on time. Only few people who can deliver good results should be employed so as to reduce overhead cost. Hence, with a thorough job description and evaluation for such staff, the fish farm can yield optimum output with reduced expenditure on variable inputs.

The management practices on the farm are other determinants of a successful project. In an established fish farm the following are to be carefully assessed and managed: the water quality, stocking density, feeding, the stocked fishes, records, environment, disease control etc. All these factors interplay to ensure the success of the fish farm project.

1. **Water Quality Assessment**

Water is the only medium where fishes can live and grow. Without water, any fish farm project is impossible. Therefore, the quality of water in the farm is very essential to the success of the fish farm project. If the water is not properly managed, it will be unfavorable to the fish which might lead to mortalities due to depletion of oxygen required by fish for survival, increase in the ammonia and nitrite contents of the water, outbreak of diseases etc. Hence, for proper water management, the following water parameters must be monitored regularly;

1. **pH**

This is a measure of the alkaline and acid content of water given by the level of hydrogen ion (H+) dissolved in the water. pH (potential of Hydrogen) is measured with the pH meter graduated from 1 to 14 with 7 as a measure of neutrality, while 1 – 6.9 is the measure of water acidity, and 7.1 – 14 is the measure of water alkalinity.

Fig. 3.1.5: Digital pH Meter

If the water is too acidic for the survival of fish, lime is added to raise its alkaline level and thereby reduce the acidity. Lime provides a buffer to pond water against pH fluctuations. The limes available in Nigeria include; Agricultural limestone (calcium carbonate), slaked lime (calcium hydroxide); quick lime (calcium oxide); and calcium cyanide.

If the water is too alkaline, $AlSO_4$(alum) can be used to lower it as it decreases the total concentration of basic minerals in the water. Thus, the ideal pH for most culturable fish species is between 6.5 – 8.0, and the constant monitoring of this parameter will provide a conducive medium for the survival of the fish.

(ii) **Dissolved Oxygen**

Fish, like all other animals require oxygen for respiration (breathing). Thus, oxygen is very important to the survival of fish in the water. But because oxygen required by fish should be exchanged through the gills (organs of respiration in fishes), it must be in dissolved form in water. Hence, the amount of dissolved oxygen in water must be enough for the survival of the fish, from fry stage to grow out stage. The ideal level of dissolved oxygen is between 5mg/litre – 8mg/litre of water for rearing of fish fries and higher for grow-out fishes. However, if the level of dissolved oxygen is low, aerators can be used in the case of fish fry rearing, and regular change of water is done in grow out ponds. Also, powerful aerators can be used in grow out fish tanks. When the level of dissolved oxygen in water is low, fish get stressed up and tend to come close to the water surface for air by positioning erect, head up, usually called "hanging". This should be corrected immediately by changing the water in the fish tank and aerating it by inflow of fresh water.

(iii) **Temperature**

High temperature is unfavorable for fish culture; therefore, care should be taken to always keep the water temperature at a moderate level of between 27°c – 30°c. Lower temperature tending towards cold water or freezing water temperature must be avoided especially in regions like Jos (Plateau State) in Nigeria. A thermometer is used in measuring pond water temperature.

(iv) **Ammonia and Nitrites**

Ammonia and nitrites are toxic compounds that are harmful to fish and must be controlled regularly in fish ponds. The levels of ammonia and nitrites when increased by the level of decomposed unfed feeds at the bottom of fish tanks lead to depletion of dissolved oxygen in water which becomes harmful to fish and can lead to mortalities. When there is increased level of ammonia and nitrites in fish pond water, the fish get stressed up and hang at the water surface. In such conditions the water must be changed immediately and fresh water impounded into the fish tank, quality feed are given to the fish and water temperature kept at moderate level.

(v) **Carbon (iv) oxide**

Carbon (iv) oxide is to plants what oxygen is to animals. It is the end product of animal respiration, fishes inclusive. However, it is a major component of plant respiration and must be expelled by animals. Too much of CO_2 in the water makes respiration difficult for fish, as there will be less oxygen in the water. Such situation must be controlled quickly by aeration of the water using power air-pumps.

(vi) **Turbidity**

This is the measure of suspended matter in the pond water. It is denoted by absence of transparency in the water. Freshwater and brackish water are normally transparent, colourless, and tasteless while marine water is normally salty. Turbidity is measured using the secchi disc, and it should measure less than 30cm for the water to be described as conducive for fish culture. The more turbid pond water becomes, the less visual it becomes, and the more difficult it is for light penetration. Hence, the visual colour of pond water should be watched closely, as heavily silted or muddy water due to decomposed feed accumulation is known to cause gill clogging in fish.

If pond water is too turbid, application of fertilizers or fish feeds should be suspended and water drained off to about 50%

after which fresh water is impounded. This will allow the fish coming to the surface (due to stress) to become agile and live and grow normally. Turbidity generally is caused by the accumulation of silt of uneaten feed in concrete fish tanks; and can be controlled by regular change of water in the fish tanks.

Other relevant water parameters in fish culture are nitrate, alkalinity, hardness, Hydrogen sulfide, conductivity, salinity etc

The table below shows some important water parameters with their required ranges for fish culture.

Table 3.4 Pond water parameters with their required ranges

S/N	PARAMETER	OPTIMUM RANGE
1.	pH	6.5 – 8.0
2.	Dissolved Oxygen	5mg/l – 8mg/l
3.	Turbidity	30cm
4.	Temperature	27°C – 30°C
5.	Alkalinity	20mg/l – 150mg/l
6.	Hardness	20mg/l – 150mg/l
7.	Ammonia	omg/l
8.	Nitrite	omg/l

I. Determination of Stocking Density

Before stocking fish in a pond, the stocking density of the pond must be determined. Recent researches have shown that overcrowding of fish in ponds results in cannibalism, stiff competition for food and feed, easy spread of disease outbreak, disease transfer, stunted growth etc. Therefore, the need for normal stocking cannot be over emphasized if the fish farmer desires good fish growth and reduced likelihood of cannibalism and diseases.

The stocking density is dependent on the farmer's ability to manage the stocked fishes and it varies with culture systems and fish sizes. However, for catfish being one of the most cultured fish, certain densities act as guide (although they can be manipulated depending on the farmer's management skills). These are:

1. 40 – 60 fish/cubic metre for concrete and other mobile tanks.
2. 60 – 80 fish/cubic metre for earthen ponds.

The above figures act as guide to the farmer in determining the number of fish that can be raised or produced in his ponds. The density may be higher at stocking, but it should be noted that fish growth will necessarily result in demand for more space. This can be manipulated back and forth depending on the management ability of the farmer.

In recirculatory systems however, the stocking density is always higher; and it can be between 200 – 300 fish/cubic metre (being an intensive culture system).

To practically calculate the number of fish that can be stocked in a particular pond, the following steps are to be taken;

1. Take the measurement of the pond and get the dimension in length, breadth and depth (in metres)
2. Multiply the figures obtained and your result will be in m³(cubic metre).
3. Then, multiply the answer by the number of fish required to be stocked per cubic metres (from the stocking densities stated above).
4. The answer is the number of fish to be stocked in that pond.

For example, in a concrete tank of dimension (8m x 5m x

1.5m), which is equivalent to 60m³, you multiply 60m³ x 40 fish per cubic metre to get 2,400 fish which will be stocked.

If using 60 fish/cubic metre, it will be 60m³ x 60 fish = 3,600 fish which will be stocked.

The same goes for other culture systems.

Also, in a polyculture project e.g. rearing catfish and tilapia in the same tank, care must be taken to determine the stocking ratio of such species which depend on certain factors. More often than not, catfish and tilapia are stocked at ratio 2:1 minimum or ratio 5:1 maximum. But the feeding habits of such species must be taken into consideration before stocking to avoid high level of cannibalism and competition for food at the same level of water.

1. **Fish Nutrition**

Cultured fish derive their daily nutritional requirement in two main ways; they feed on natural foods occurring naturally in the ponds such as phytoplankton, zooplankton, worms, insect larvae, small plants and other plant matters and they also feed on artificially formulated or compounded feed made up of ingredients of varying sources and proportions.

To grow to the required size within a short period of time, it is required that fish be supplied with the required nutrient at the right time. Catfish for instance is a carnivorous fish, so a lot of protein is required in its feed. When fed adequately, catfish fingerlings should grow to an average weight of one kilogram within six months. This will only be achieved when feed with a crude protein content of 40 – 48% is fed to the fish in the right quantity (proportion) and at the right time.

1. **Feeding Rate**

The quantity of feed that should be fed to a fish at different stages of life depends on its body weight. A certain percentage of the total weight of the fish population is used to calculate how much feed should be given at a time. Feeding rate ranges from 5.5% to less than 1% of the body weight.

For example, a population of one thousand juveniles each with an average body weight of 10 grams should be fed at 5.5% of the body weight daily. The total weight of the whole fish population will be;

10g x 1000 pieces =10,000g

The feeding rate at 10g =5.5%

Therefore, 5.5/100 X 10,000g

=550g

This means, 550g of feed (the right grade) will be given daily to the whole fish population at that stage.

Table 3.5Feeding rates at various fish sizes

S/N	SIZE OF FISH	FEEDING RATE
1	5g– 10g	5.5 – 6%
2	10g– 50g	4.5 – 5.5%
3	50g– 100g	4.0 – 4.5%
4	100g– 250g	3.0 – 4.0%
5	250g– 500g	2.0 – 3.0%
6	500g– 750g	1.5 – 2.0%
7	750g– 1kg	1.1 – 1.5%
8	1kg – 1.2kg	0.9 – 1.1%
9	1.2kg – 1.5kg	0.8 – 1.0%
10	1.5kg – 2kg	0.7 – 0.9%

2. **Fish Feed Ingredients**

Fish feed ingredients are classified as macro and micro ingredients depending on the type of nutrient they supply. The macro ingredients are needed in large quantities; they form about 98% of the feed while the micro ingredients form only about 2 – 3% of the whole formulation. Macro ingredients are protein (fish meal, soya bean meal, groundnut cake, blood meal, palm kernel cake etc which have above 18% crude protein) and carbohydrate or energy (maize, wheat, millet, sorghum etc which have less than 18% crude protein) supplying ingredients. They could be of either plant or animal source. The micro ingredients supply trace elements like vitamins, calcium, phosphorus and amino acid. Examples of micro ingredients commonly used in fish feed are lysine, methionine, vitamin C, sodium chloride, bone meal, oyster shell meal etc. Both the micro and macro ingredients are mixed together thoroughly after being ground into powder to form a whole mix.

3. **Feed Conversion Ratio (FCR)**

This is the amount of feed needed to produce one kilogram of flesh. It is a measure of how efficient a feed is. The smaller the ratio, the more efficient the feed is i.e. a feed with a FCR of 0.5 will be more efficient than a feed with a FCR of 1.2. In catfish, highly proteinous feeds usually have a low FCR, so only small quantities of such feed will be needed to produce a unit increase in flesh.

FCR = Total feed fed (kg)/ Increase in weight (kg)

A farmer's real cost of feeding is calculated as FCR x cost/kg of feed.

4. **Feeding Regime**

   Fish should be fed in small bits to enhance digestibility and reduce feed wastage. When the total daily feed quantity has been calculated using the appropriate rate, this should then be divided into three or four bits and given to the fish at three or four hourly intervals.

   By doing this, the fish will be hungry and ready to eat at each regime. This allows more feed to be converted to flesh. The feeding regime may for instance be put at 8am, 12pm, 4pm, 8pm or whichever the farmer chooses.

5. **Fish Feeding**

   This is a daily exercise which involves feeding the various sizes of fish on the farm with their respective feed grades at regular intervals. The feeds come in grades for easy ingestion and absorption by the fishes, considering the fact that their mouth sizes vary with their age. Two feed types are commonly used in fish farming, the floating (extruded) feeds and the sinking (pelletized) feeds. These feeds have varying grades from starters to finisher, and they are available in commercial quantities as *0.2mm, 0.3mm – 0.5mm, 0.5mm – 0.8mm, 0.8mm – 1.2mm, 1.2mm – 1.5mm, 2mm, 3mm, 4.5mm, 6mm and 9mm* with little differences in description depending on the feed manufacturers.

   These are fed to the fish at different stages of their lives. For instance, the 0.2mm to 1.5mm grades(starters) are used in the hatchery and nursery for fish seeds; while from 8 weeks old (as juveniles) the feeds can be switched to 2mm, and as they grow on, the feed grades are switched until they become adults and can take 9mm (finisher).

Such varying sizes of feed are made possible through the engagement of different discs in the production process. The discs which are connected to the pelletizing machines produce the feeds in the various grades as described above.

The progression of feed grade switching is from 2mm (for juveniles) – 3mm (for post juveniles) – 4.5mm (as grow outs) – 6.0mm (as sub adults) – 9mm (as adults). The broodstocks are to be fed as adults i.e. with 9mm (finisher) or any other specially formulated feed as the case may be. With this in place, the fish should reach table size within a short period (provided other growth factors are in place).

2. **Fish Sorting, Grading and Sampling**

These are techniques used in sizing fishes and monitoring their growth rate, feed efficiency and health status. In the hatchery and nursery, sorting and grading are done to discourage cannibalism and prevent stiff competition for feed (by grading the fishes into relatively uniform sizes). But in grow out (production) ponds, the fishes are sorted and graded i.e. sized into divisions, after which they are sampled so as to evaluate the efficiency of the feed on their weights as well as have a clear view/access to the entire population, observe their health status among other things, which will suggest whether they are to be treated, or transferred. Sorting and grading are also done when sales are to be made; for instance, one may sort out fishes that have already attained a target weight from those yet to reach the weight. This exercise also assists in discovering the fish that are not growing (runts) which will be separated from the well growing ones for proper concentration, while they (the runts) can be processed to smoked fishes or disposed to prevent feed wastage and thus minimize cost of feeding.

In sampling, you take out the entire population (if few) and weigh them or you take a good representative percentage of the population (if many) and get their total weight. You then divide this total weight by the number of fish taken out to get their average weight.

For example;

If 475 pieces of fish are taken out and weighed to be 92kg $\equiv$ 92,000g;

The average body weight per fish =92,000g/475pcs

= 193.6 $\approx$ 194g

If the total population is 1000 fishes, the average weight will be 194g X 1000 fishes = 194,000g

At 3% feeding rate (from Table 3.5), we have,

3/100X194,000g

=0.03 x 194,000g= 5820g $\equiv$ 5.82kg

That is, approximately 6kg of feed is to be given to the total fish population on a daily basis (which can be spread into regimes). This sampling is carried out regularly to know how the fish are performing and when to increase the quantity of feed as a result of growth.

3. **General Maintenance**

   In a good fish farm, management practices revolve around the maintenance of equipment used on the farm, from hatchery equipment to even the water pump in the well/borehole. General maintenance of the fish farm environment like grass cutting, cleaning of ponds regularly, changing of water as and when due, disinfecting trees, tanks etc are very important as this will make it impossible for some hosts of disease pathogens to thrive within the environment, thereby increasing the chances of fish survival.

   Fish farming is actually an interesting practice and a very green area for investors. But as already highlighted in this chapter, one will have to know what it takes and do it the right way so as to get the expected results. As this chapter is being concluded, it is worthy of note that all the factors already explained interplay to make any fish farming project a success; the absence of any of these factors may lead to failure.

Study Questions:

1. Define *aquaculture.*
2. Write short notes on the following:

1. Location feasibility
2. Biotechnical feasibility
3. Economic feasibility.

1. Highlight four (4) water parameters relevant to fish culture with their respective optimum (suitable) levels or ranges.
2. What is the function of Secchi disc?
3. A fish farmer intends to stock his newly constructed pond of dimension 16ft X 12ft X 5ft with catfish juveniles. If he decides to stock at a stocking rate of 41fish/cubic metre, how many juveniles will he stock in the pond?
4. List five (5) energy yielding and four (4) protein yielding fish feed ingredients.
5. Why should sorting and grading be carried out in fish production?

Chapter Four

FISH DISEASES AND MANAGEMENT

4.1 Disease Defined

A disease can be defined as an unwholesome condition manifested by the departure of the body from the normal healthy state to that of discomfort, sickness and death. Technically, it refers to a complex interaction between a susceptible host, a pathogen and the environment. With the presence of pathogen in effective number, a susceptible host suffers an infection if the environment is adverse to it.

4.2 Causes of Fish Diseases

Fish may die as a result of a disease or pollution of the environment. Diseases affect a particular fish species or a particular age group in fish populations while pollution destroys all fish populations irrespective of age or species. The major disease promoting factors in fish ponds include;

1. Accumulation of decaying organic matter from excess feeding, fertilization, and faeces which provide ideal medium for harmful micro-organisms to grow and multiply.
2. High stocking density and overcrowding bring about stress condition to fish which lead to immunity deficiency.

3. High temperature in stagnant water such as fish ponds provides harbor for the growth of micro-organisms.
4. Poor diets, irregular and inadequate feeding enhance disease transmission and outbreaks in ponds.
5. Entrance or introduction of infected wild fish stock into the pond.
6. Exit of contaminated or polluted water into the pond.
7. Presence of fish parasites in ponds and
8. Prolonged handling stress.

4.3 Symptoms of Fish Diseases

These refer to the changes observed in the body or behavior of the fish indicating a sign of disease condition. The following symptoms when noticed in the fish stock indicate the presence of diseases:

1. Loss of appetite (cessation of feeding)
2. Weak, slow or erratic movement
3. Appearance of blood spots on some parts of the body (Haemorrhagic condition)
4. Presence of wounds on the skin (ulceration and necrosis)
5. Frayed or chopped fins and tail
6. Presence of boil like structures on the body (lesions)
7. Appearance of pale, bloody and puffy gills
8. Skin discoloration and patches on the skin
9. Bulging of the eyes
10. Swellings, nodules, scars and blisters
11. Jumping and flashing
12. Loss of weight
13. Poor or stunted growth
14. Bone deformations
15. Gasping at the water surface (hanging)

16. Spiral movement
17. Loss of balance
18. Crowding at the water inlet or outlet
19. Belly burst
20. Shrinking or cutting barbells (in catfish)
21. Mass mortality in the pond etc

4.4 Types of Fish Diseases

Diseases of fish are broadly classified into two main groups as follows;

1. Infectious diseases and
2. Non-infectious diseases.

4.4.1 Infectious Diseases

These are diseases caused by pathogenic organisms present in the environment or carried by other fish species. They are broadly classified as parasitic, bacterial, viral and fungal diseases. Their mode of transmission is usually from host to host. Pathogens from infected fish are released into the water from where they attack other fish through the oral openings, wounds and skin. In some cases, transmission is from parent stocks to their offsprings (transovarial).

4.4.1.1 Parasitic Diseases

1. **Fish Tapeworm (Cestode)**

- Causative Organisms – *Corallobothrium fimbriatum, Ligula intestinalis, Diphylobothrium latum.*
- Clinical Signs (Symptoms) – parasite attaches itself to the intestinal walls of the host; and though the affected fish shows no outward indication, it may be listless, lose weight or become sterile.

- Contributing Factors – use of brood fish infested with tapeworm, purchase of contaminated fry and fingerlings, and the use of surface water containing tapeworm infested hosts. The droppings of fish-eating birds in and near the pond can also introduce tapeworms.
- Prevention and Treatment – fish farmers are to avoid maintaining or purchasing infected fry, fingerlings or broodstock; there should be draining, drying and disinfection of ponds between fish crops in order to eliminate or reduce exposure to intermediate hosts. Currently, there are no therapeutic agents available.

1. **Ichthyopthiriasis (Protozoan)**

It is commonly called **ich** or **white spot disease**. All freshwater fish are susceptible, and it is also a disease of aquarium and hatchery reared fish.

- Causative Organism – *Ichthyopthiriasis multifilis* (a ciliated protozoan).
- Clinical Signs (Symptoms) – parasites appear as small raised spots that resemble sprinkled table salt over the entire body surface, particularly on the fins. The affected fish may also be seen flashing against the bottom or sides of the tank and heavily infected fish often congregate at the inlet or outlet of the pond or tank.
- Contributing Factors – poor water quality, malnutrition, increased susceptibility when the water source is contaminated with wild fish as well as when water temperature is 16°C to 24°C. Microscopic examination verifies the presence of ich protozoan.
- Prevention and Treatment – contaminated water supply, nets and other equipment must be avoided, good water quality

must be provided, and nutritionally adequate feeds should be offered. Formalin, table salt, copper sulphate and potassium permanganate can be used as therapeutic agents.

I. **Fish Lice (Crustacean)**

Fish lice are parasitic Branchiurans of the genus Argulus. They are related to the anchor parasite as they attach themselves to the skin by suckers. All fresh water fish are susceptible.

- Causative Organism – *Argulus sp*
- Clinical Signs (Symptoms) – infected fish will flash or rub against the tank, pond bottoms or sides. Also, they will be listless and show red spots; and when infections are heavy, the fish start dying.
- Contributing Factors – stocking of lice-contaminated fish into parasite-free populations allows the lice to spread. And depending on size, examination by eye or microscope verifies the presence of fish lice.
- Prevention and Treatment – stocking of parasite-free fish is the best prevention. However, Masoten (Dylox) can be used as a therapeutic agent.

I. **Monogenetic Fluke**

- Causative Organisms – Gyrodactylus, which is usually found on the skin and the fins; and Dactylogyrus, which attaches to the skin.
- Clinical Signs (Symptoms) – fresh water fish infested with skin-inhabiting fluke become lethargic, swim near the surface, seek the sides of the pond and their appetite dwindles. Also, they may be rubbing against the bottom or the sides of the

holding facility (flashing); the skin where the flukes are attached show areas of scale loss and may ooze a pinkish serous fluid. Heavy gill infestations result in respiratory disease such that gills may become swollen and pale, and respiratory rate may be increased leading to fish becoming less tolerant to low oxygen conditions. "Piping", gulping air at the water surface may be observed in fish with severe respiratory diseases.

Large numbers of monogenetic fluke on either the skin or gills may result in significant damage and mortality. Secondary infection by bacteria and fungus is common on tissue that has been damaged by monogeans.

- Treatment – salt treatment (the right concentration), and Formalin (35% - 37%) which has been most effective.

Other parasitic diseases of fish include **Costiasis (Protozoan)** caused by *Costia necatrix*, **Anchor worm (Crustacean)** caused by *Lernea sp.*, **Yellow grub (Trematode)** caused by *Clinostomium marginatum*, and **Leeches (Annelid)** caused by *Piscicola geometrica*.

4.4.1.2 Mycotic Diseases(Fungal)

I. **Saprolegniasis**

- Causative Organisms – *Saprolegnia parasitica* which usually attacks adults, fry and eggs of fish. All freshwater fish can be affected.
- Clinical Signs (Symptoms) – fish have a general cotton-like or fur-like appearance usually associated with localized discoloured areas or lesions. Fungus assumes the colour of materials suspended in the water.

- Contributing Factors – fungal infections are generally secondary and indicate other adverse conditions; fungal infections seldom become established on healthy fish unless they have been subjected to stress or injury; stress conditions include prolonged periods of very low temperatures, malnutrition and possibly low dissolved oxygen; fungi infections are also a sign of bullying by other fish (e.g. fin nippers will damage the fins of other fish); microscopic examination verifies the presence of fungi.
- Prevention and Treatment – maintain good water quality and feed nutritionally adequate feeds throughout the year; feeding just before winter and in early spring is very important. Copper sulphate and potassium permanganate can be used as therapeutic agents, and Formalin has also been very effective.

2. Cataracts

These are fungal growths indicated by white or gray material covering the eyes only. Treatment with any aquarium fungicide should work; and during therapy, attention should be made to ensure that ammonia and nitrite levels are kept within acceptable measures. The probability of this condition increases with water rich in ammonia and nitrite.

Another fungal disease is **Branchiomycosis** which leads to blockage of gills capillaries. It is caused by *Branchiomyces sp.*

4.4.1.3 Bacterial Diseases

I. Bacteremia (Haemorrhagic septicaemia)

- Causative Organisms – Bacteria like *Aeromonas hydrophilis or Pseudomonas fluorescens* and possibly other bacteria cause bacteremia (bacteria in the blood). All fish can be affected.

- Clinical Signs (Symptoms) – infected fish are listless and lethargic with shallow, irregular-margined reddish sores or ulcers on the sides. They are popeyed, and have enlarged (swollen) fluid-filled belly. Raised scales and red streaks in the fin rays and at the bases of the fins are also symptoms of this disease as well as reddened areas around the anus.

- Contributing Factors – outbreaks may occur in spring when the water warms particularly when the fish spawn, are handled, moved or are overcrowded. Outbreaks also occur when dissolved oxygen content of the water, and possibly when other conditions such as malnutrition and diseases weaken the fish. Laboratory culture confirms bacteremia.

- Prevention and Treatment – precautions include avoiding rough handling or overcrowding especially during summer, maintaining good water quality and providing a well-fortified feed containing greater than recommended levels of ascorbic acid (Vitamin C). Possible therapeutic agents include antibiotics like oxytetracycline (terramycin) in the diet; the addition of oxytetracycline to fish transport water may also retard the transfer of the bacterium but will not cure already infected fish.

2. Columnaris

- Causative Organisms – *Flexibacter columnaris* sometimes called *Cytophaga columnaris*. It is a serious disease of young salmonids, catfish and many other fish; it affects the fingerlings and juveniles of the African catfish (saddle back disease in catfish fingerlings and juveniles). All fish are susceptible.

- Clinical Signs (Symptoms) – affected fish show discoloured patches on the body with little or no haemorrhaging or sloughing of scales; discoloured patches and scales loss superficially resemble damage caused by fungus infections; dermatitis on the dorsum(usually bilateral) is pathognomonic. Other signs include mouth and barbel erosion, fin erosion, tail loss and decayed areas in gills. Mortality ranges from 30-70%, but can be higher in untreated cases. Condition can be systemic with high mortality.

- Contributing Factors – mechanical injury caused by rough handling, especially when water temperature exceeds 20°C can contribute to an infection. It is very common in hot season as outbreak is temperature related, hence the endemic nature of the disease. Overcrowding in holding and transport facilities, poor water quality (low dissolved oxygen), fluctuating water temperatures, and malnutrition also contribute to this condition. Microscopic examination and laboratory culture verify the presence of columnaris.

- Prevention and Treatment – as in Bacteremia, possible therapeutic agents include water treatments with potassium permanganate or Diquat. The medication of feed with oxytetracycline may be helpful if the infection is systemic (internal); usually in Nigeria, Tetracycline is the drug of choice. Topical application of potassium permanganate or acriflavine is recommended.

3. Enteric Septicaemia of Catfish

- Causative Organisms – *Edwardsiella ictaluri*. Because of one of the signs of the diseases, it is sometimes called **"hole-in-the-head"** disease. It affects channel catfish (primarily fingerlings and yearling catfish).

- Clinical Signs (Symptoms) – fish may have a "**hole-in-the-head**" lesion between the eyes. This may appear as a white or reddish raised area before the hole appears. Fish will also have pimple-like lesions over the general body surface, a typical dropsy appearance; bloody looking internal organs or yellowish or reddish fluid in the body cavity and fish may also cease feeding, become listless, hang tail-down in the water, but spasmodically swim rapidly in circles.

- Contributing Factors – low dissolved oxygen, high ammonia and nitrite concentrations, and water temperatures between 21°C and 27.8°C appear to be the conditions usually associated with the outset of the disease. Laboratory culture verifies the existence of the bacteria.

- Prevention and Treatment – good water quality should be maintained, dissolved oxygen should be kept above 4ppm, and good quality feed containing supplemental ascorbic acid (Vitamin C) should be served. Possible therapeutic agents include antibiotics such as oxytetracycline in the feed.

4. Corynebacteriosis

This is usually characterized by bulging eyes. Corynebacteria causes swelling in the head which pushes the eyes outward; it is caused by overcrowding, poor water quality and excess of ammonia and/or nitrites.

5. Myxobacteriosis

This infection is rather uncommon but fairly easy to treat. Its symptoms may include black patches on the body and fins, and the body may become bloated or swollen in some areas. Its probability is intensified by overcrowding and poor water quality with high levels of ammonia and/or nitrite.

There is only one medication designed specifically for myxobacteriosis – Phenocide (by Aquatronics).

Other bacterial diseases of fish include **Frunculosis** caused by *Aeromonas sp*, **Fish Tuberculosis** caused by *Mycobacterium sp*, **Fin rot** caused by *Cytophaga psychrophilia*, and **Bacterial gill disease** caused by *Marimum filamentous*.

4.4.1.4 Viral Diseases

1. **Viral Haemorrhagic Septicaemia**
   This is caused by *Rhabdovirus sp*, with ulcerative haemorrhagic dermatitis as its symptom.

2. **Carp Spring Virema**
   This is caused by *Rhabdovirus carpio*, with dark skin, loss of balance and exophthalmia as symptoms.

3. **Infectious Haemorrhagic Necrosis**

   This is caused by *Reovirus sp*, with loss of balance and accumulation of fluids in the body cavity as symptoms.

4.4.2 Non - Infectious Diseases

These are diseases caused by environmental problems, nutritional deficiencies or genetic defects; and they are broadly classified according to the causative factor. They are not contagious and usually cannot be cured by medications. Environmental diseases are the most important in commercial aquaculture, and they include low dissolved oxygen, high ammonia, high nitrite and natural or man-made toxins in the aquatic environment. Proper techniques of managing water quality enable producers to prevent most environmental diseases.

4.4.2.1 Nutritional Diseases

These are diseases caused by dietary deficiencies (unbalanced diet), improper processing and storage of feed and feed ingredients as well as feed contamination (chemical or biological). A balanced feed consists of all the six classes of food in adequate proportions (carbohydrate, fats and oil, protein, mineral salts, vitamin and water). Shortages and excesses of these components lead to nutritional diseases. Examples of such nutritional diseases are given in Table 4.1 below.

Table 4.1: Nutritional Diseases of Fish

S/N	DISEASE	DEFICIENCY	SYMPTOMS
1.	Goitre	Iodine	Swelling between base of tongue and last gill arch.
2.	Broken Head	Vitamin C	Skull inflammation and breakage.
3.	Anaemia	Iron	Weakness and lethargy.
4.	Bone deformity	Calcium Phosphorus	Curvature of bone.
5.	Eye cataract	Vitamins A and B Methionine	Eye deformities.
6.	Exophthalmia	Vitamins A and B Folic acid	Bulging of the eyes.
7.	Fin and skin haemorrhages	Vitamins A, C and K	Blood spots on fins and skin.
8.	Fin erosion	Vitamin C and Essential Amino acids	Frayed or chopped fins.
9.	Fatty liver	Essential Fatty acids and Folic acid	Inflammation of the liver.
10.	No blood disease	Folic acid	Anaemia and death.

4.4.2.2 Environmental Diseases

These are diseases caused by changes in the pond environment. Causes of these diseases include:

1. Water pollution (agricultural or industrial)
2. Application of pesticides and fertilizer
3. Oxygen depletion/saturation
4. Excess level of ammonia, hydrogen sulphide etc
5. pH and temperature variation
6. Presence of toxic substances in pond water.

Examples of such environmental diseases are given in Table 4.2 below.

Table 4.2: Environmental Diseases of Fish

S/N	Disease	Cause	Symptoms
1.	Gas Bubble Disease	Oversaturation of gases such as oxygen and nitrogen	Appearance of tiny bubbles under the skin of fish.
2.	Brown Blood	High level of nitrogen and hydrogen sulphide in ponds	Brown gills and chocolate brown blood.
3.	Algal toxins	Excess bloom of blue-green algae in ponds (*Coelophaerium dubium*)	Mass mortality of fish in ponds.

4.4.2.3 Constitutional Abnormalities (Neoplastic Diseases)

These diseases refer to the abnormal growth or tumor that may appear in any part of the body with loss of structural and functional ability of the affected organ.

Examples include **Pappilomas**, a wart-like tumour found on the skin, lips, fins and opercula of the fish and **Viscera**

Granuloma, an inflammation/tumour of the visceral organs such as the liver, stomach, kidney, heart and gills.

4.5 Economic Importance of Fish Diseases

Diseases outbreak in ponds is more common in intensive fish culture where large number of fish is cultivated in a relatively small volume of water. It is also capital intensive where a lot of financial commitment and labour are involved. However, in the event of an epidemic quite a number of side effects become inevitable. These include:

1. Economic losses due to high cost of drugs, loss of fish stock, and diseased fish are generally unattractive to consumers thereby attracting less market value.
2. Excessive use of drugs and chemicals in ponds may lead to pollution of the environment.
3. Parasites and pathogens may develop resistance due to prolonged usage of drugs.
4. Some diseases are transferred to the consumers e.g. Fish tapeworm.
5. Diseases generally retard growth, fecundity and reproduction hence affecting fish population.
6. Fish survivors of an epidemic serve as reservoirs of the disease and may transmit the disease to other fishes.

4.6 Prevention of Fish Diseases

It has been said that prevention is better than cure; and for fish, this is best achieved through proper health management practices. These include;

1. Proper feeding of fish with adequate quantity of good quality feed.
2. Handling must be done with adequate care to minimize stress.

3. Fish for stocking must be obtained from disease free zones and should be treated for ecto-parasites before stocking.

4. Overcrowding should be avoided as much as possible. Correct stocking density must be adhered to. This however depends on the culture system being operated.

5. Entry of wild fish and other unwanted aquatic animals (frogs, snails, birds etc) should be avoided. Water inlets must be properly screened while pond dykes should be raised or fenced with fine netting materials.

6. Dead fish must be removed and sick fish quarantined immediately.

7. Water quality parameters must be maintained at optimum levels (dissolved oxygen, pH, temperature etc).

8. Regular disinfection of appliances and hatchery equipment should be carried out to destroy harmful microorganisms.

9. Broodstocks must be removed from the young fish as they may serve as reservoirs for certain diseases.

10. Prophylactic use of drugs such as antibiotics must be done with care, and correct dosage must be administered.

11. Immunization has become one of the most important ways of preventing communicable diseases in all animals including fish.

12. Genetic manipulation, involving the crossing of two different fish species with high degree of resistance to produce a hybrid that is resistant to diseases.

13. Fish farmers are advised to contact the nearest fishery extension agent in the event of disease outbreak in farms.

4.7 Control of Fish Diseases

Fish diseases can be controlled through different methods. These include;

1. Test and Slaughter – which involves the destruction of all the

affected fish or the whole population by burying or burning them.

2. Quarantine and Restriction of Movement – which involves the transfer of infected fish to an uninfected area under close observation. However, this has to be done for a period not shorter than the incubation period of the suspected disease.

3. Therapy – which involves the use of approved drugs and chemicals to destroy the pathogen. It is done in two ways:

1. External Treatment

1. Bath – immersion of the infected fish into water soluble compounds of low concentration for a long time or high concentration for a short time.

2. Swab – application of high concentrated solution to the affected part (tissue or organ) for a short period of time.

1. Internal Treatment

1. Treatment via Diet – drugs are incorporated into fish feed, though this is not very effective due to loss of appetite.

2. Injection – injection of drugs like oxytetracure injection 20%LA intramuscularly using sterilized syringes and hypodermic needles. This is highly effective, though not in common practice.

1. Disruption of Parasites' Life Cycles – which involves the destruction or reduction of a link in the transmission cycle of animal parasites to control infectious diseases of fish.

2. Diagnosis of Disease caused by Feed – in doing this, environmental effects, primary infections and parasitic infestations must be ruled out at first. If this has been reliably done, feed-caused disease can be suspected. The assumptions concerning the cause of the disease can be confirmed by changing feed. Af-

ter feeding with good quality and biologically complete feed, the fish should recover.

Study Questions:

1. Briefly explain the term *disease*.
2. List five (5) disease promoting factors in fish ponds.
3. Highlight twelve (12) symptoms of fish disease.
4. Give three (3) examples of parasitic disease in fish.
5. Mention six (6) nutritional diseases of fish.
6. What are *neoplastic diseases*?
7. State three (3) ways by which fish diseases can be controlled.

Chapter Five

FISHING EQUIPMENT

5.1 Definition

Fishing equipment refer to the various tools, materials, craft and gear engaged in fisheries ranging from those used in open seas (capture fisheries) to those used in aquaculture (culture fisheries). They come in different forms and sizes, each adapted for its specific purpose.

Some of these equipment are engaged at the pre-production stage, some engaged in the course of the fish production process while others only become useful at harvest.

5.2 Fishing Gear and Craft

The use of craft and gear in fishing technology plays very important roles and helps in enhancing commercial based production. The success of fishing largely depends on how and which types of nets are used to capture the fish.

There are two main types of devices used to capture fishes in both marine and inland fisheries:

1. Nets or gear — these are instruments used for catching fish.
2. Craft or Boats — which provide platform for fishing operations, carrying the crew and fishing gear.

There are various types of gear and craft used in different parts depending upon the nature of water bodies, the age of fish and their species. Some nets are used without craft; however, others are used with the help of craft. Generally, locally made gear and craft may be non-mechanized or mechanized.

5.2.1 Craft and Boats:

There are many types of fishing craft being successfully made and used for marine and inland fisheries.

1. **Marine Fishing Craft**

   Different craft are used due to different conditions of sea on the east and west coasts.

   Craft used on the East Coasts include Catamaran, Masula boat, Nauka and Dinghi, Tuticorin boats or Fishing luggers etc

   Craft used on West Coasts include Dugout canoes, Plank-Built canoes, Outrigger canoes, Built-up boats, Coracle, Shoe Dhonie, Kakinada Nava etc

2. **Marine Fishing Gear**

   Various types of gear are used for fishing in sea. They may be of different sizes, shapes and designs. These gear may be made by fishermen. They are also manufactured in cottage industries. The most commonly and widely used fishing gear are different types of nets. They are used for catching large fishes offshore. The main type of nets being used are boat seine, shore seine, bag nets, fixed or stationary nets, drag nets, drift nets and cast nets.

   1. **Seines**

      These are specially designed and large fishing nets. They are generally used in running water. When they are spread in sea;

they collect large numbers of fishes. Seines are rectangular in shape mounted on wire. They are spread vertically in the water. Seines are of two types, boat and shore seines.

2. **Trap Nets**

They are generally used for fishing in shallow waters. Trap nets are strong and made in various shapes and sizes. These nets may be stationary or fixed. Its lower part is cylindrical while upper part is conical. Interior region of the net contains one or two cone-shaped necks to prevent escape of fish. Large trap nets are called a pound net, which has a chamber with a wide gate.

3. **Drop Net**

It is square in shape and mounted with supple loops at the corners that tied in a cross at the top and is attached to a pole. Drop net is operated with a boat. It is dropped and pulled to catch fishes.

4. **Cast Net**

It is a circular and cone shaped net. It is spread from the edges of water. Its circumference is attached to lead line while its centre is attached with a rope. The net assumes shape of umbrella when it is spread on the water. When the net sinks to the bottom it is pulled and fishes are collected.

5. **Drift Nets and Gill Nets**

These types of nets are made by nylon materials. Gill nets are kept overnight in the water and then dragged. The fishes get entangled in the meshes. There are two types of these nets — simple and trammel nets. **Simple Gill Nets** are loosely woven nets. When spread in water, fishes get entangled in mesh. If the fishes try to escape the twine of the net get mingled in the gills of fishes. The fish is said to be gilled (captured by gills) and hence the name given 'Gill Net' while the **Trammel Gill Net** has a float line at the top and a dead line at the bottom with two walls attached to these lines. It is generally operated to catch small fishes.

6. **Fixed or Stationary Nets**

These nets are used to catch fish at inshore water during low tides. These nets are kept fixed with the help of floats, sinkers and stakes. It is rectangular or conical in shape. They are available in various sizes.

7. **Bag net**

It is conical in shape without wings. These nets are used with the help of two catamarans or boats. In the coasts of Mumbai and Gujarat a special type of bag net called 'dol' is used. It is conical with wide mouth. The mouth is fixed on a bamboo.

8. **Scoop Net or Dip Net**

It is round in shape and is used to capture delicate fishes. It is like a finger bowl and can be moved swiftly in a scooping manner, collecting the fish.

9. **Hooks and Lines**

They consist of two types of hand lines and long lines. Various types of hooks are used, such as chain hooks, baited hooks, revolving and non-revolving hooks for capture of larger fishes.

10. **Trawls**

These are large dragging type nets. There may be two types of trawls with beam called beam trawls and otter trawls.

Although these aforementioned craft and gear are mainly customized for use in the marine habitat, some of them are also useful in freshwater habitat (inland fisheries) and even in aquaculture.

Figure 5.1 below presents some common fishing gear and craft.

**Fig. 5.1: Common
fishing gears and crafts**

5.3 Maintenance of Gear

Proper care and handling of fishing gear after their use is as important as their use. Proper maintenance increases the durability of the gear. To ensure this, the following cares are necessary:

1. The gear should be washed thoroughly with clean water; and weeds, mud etc. should be removed carefully.

2. The net should then be dipped in diluted $KMnO_4$ or $CuSO_4$ or common salt solution to get rid of harmful bacteria.

3. Wash again with clean water and then spread in shade for drying.

4. To increase durability and strength of the fibre of gear, it may be kept immersed for 10-15 mins in hot tar diluted with kerosene.

Study Questions:

1. Differentiate between fishing gear and fishing craft.
2. List six (6) fishing gear that can be used in a marine habitat.
3. Highlight four (4) ways by which fishing gear can be maintained.

Chapter Six

FISH HARVESTING AND POST - HARVEST PRACTICES

6.1 Harvesting

Fish are said to be mature when they reach marketable size or weight. At this point, they are taken out (harvested) from the ponds using a variety of methods.

The fish are drained into a catch basin or harvesting trough and removed with a net. The kind of net used depends on the sizes of fish to be harvested. For instance, **Gill nets** have mesh sizes of 2-3cm and are used to harvest larger size fish usually with the intention of leaving the smaller size fishes to grow (**partial harvesting**). When the intention is to harvest all the fish in the pond, then **seine nets** which have much smaller mesh sizes may be used (**total harvesting**).

Fish may also be harvested with hook and line if only a little is needed for consumption or for recreational purposes (game fishing).

There are some other methods which are not recommended, and should never be used; they include poisoning and dynamiting, because they are dangerous and may constitute health hazards to human beings upon consumption.

6.2 Post Harvest Practices

Harvested fish are either sold fresh or stored for later sales. Also, their consumption can be as fresh or in changed form. If not to be consumed immediately, fish undergo certain practices capable of preventing their spoilage or deterioration until they are ready to be consumed. These practices are commonly referred to as **Fish Preservation** or **Fish Processing.**

Fresh fish spoil very quickly. As a matter of fact, once the fish has been caught, spoilage starts rapidly. In the high ambient temperatures of the tropics, fish will spoil within 12 hours. In a bid to prevent this, fish are conditioned in such a way that they will still be useful and hygienic beyond their fresh state.

Fish processing, which is an effective method of preserving fish from spoilage or deterioration, also changes the texture, taste and physical appearance of the fish. It literally changes the fish to a form that is more economically viable by adding value to the fish product, making the product more acceptable to the customer, maintaining the quality, improving the shelf (storage) life and even enhancing the exportation of the fish for better income to the farmer. These are however achieved through different methods such as drying, salting, filleting, freezing, frying, smoking, canning etc.

6.2.1 Drying and Salting

Drying simply implies the removal of water (moisture) from fish. The purpose is to dehydrate the fish to moisture levels at which spoilage bacteria cannot live. It is usually combined with salting.

Salting is an old fish preservation method. Fish is gutted and cleaned; then salt is packed into and on the fish. The fish is usually arranged layer by layer in a container, each layer of fish being separated by a layer of salt. The fish is then covered and left in the container for a specified period depending on the kind of fish. Next, the fish is removed and sun-dried. Sometimes cleaned and gutted fish is allowed to ferment for a few hours before being salted and dried.

in modern times however, certain drying equipment are used in the place of sun drying which makes it a more effective and time saving exercise.

6.2.2 Freezing

Freezing involves storing fishes at temperatures below -18°C. In freezing, the principle is that most micro-organisms cannot grow at such low temperatures. If fish is to be frozen, freezing must be done rapidly to avoid quality loss.

Freezing is carried out in deep freezers or specially built cold rooms. Fish may also be chilled and preserved in ice when they are only to be stored for a short while.

6.2.3 Smoking

Smoke-drying is perhaps the most widespread method of preserving fish in Africa. During smoking, fish is exposed to a combination of heat and smoke. The chemicals in the smoke inhibit microbial growth, thus aiding preservation. Through this process, fish is preserved for long periods, thus providing protein throughout the year. Also, smoking imparts a desirable taste and flavor to fish, adds value to fish and allows fish smokers to earn extra income. Smoking allows fish to be packed and transported over long distances to markets without spoilage.

Figures 6.1, 6.2 and 6.3 are examples of zero-moisture smoke-dried catfish. At this point, their shelf lives have been extended and they give certain unique tastes when consumed (added value).

Fig. 6.1: Smoke-dried catfish spread for cooling

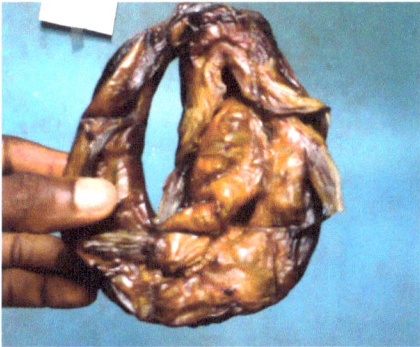

Fig. 6.2: Smoke-dried catfish ready for
consumption

Fig. 6.3: Smoke-dried catfish packaged for
marketing

Fig. 6.4: Commercial Smoking Kiln (Improved Technology)

6.2.4 Canning

This is the process where heat is used to sterilize fish packed in an airtight container. The heat destroys the food-spoilage and harmful micro-organisms and inactivates the enzymes of decomposition. The airtight container prevents gaseous exchange and prevents reinfection of sterilized fish. It is important to note that only very fresh fish should be canned since the extremely high temperatures required for the process also causes a minimal reduction in the quality of the product. Sometimes, salt, edible oil or sauce is included with the product. This process is however capital intensive and as such small scale farmers may not be able to afford it.

Study Questions:

1. Differentiate between partial and total harvesting.
2. Explain fish preservation.
3. Write short notes on the following methods of fish processing:

1. Salting
2. Filleting
3. Canning
4. Smoking

5. Freezing

Chapter Seven

FISH BREEDING

7.1 Definition

Fish breeding simply refers to the propagation of fish seeds by pairing or crossing sexually mature male fish with sexually mature female fish. It involves engaging the gametes of sexually mature male and female fish towards seeds production.

7.2 Methods of Fish Breeding

In fisheries, two main methods of breeding (spawning) exist, namely *natural* and *artificial* breeding. In the **natural**, sexually mature male and female fish are paired up in a tank of water with the expectation that as soon as the female lays the eggs, the male will fertilize them. This is actually a replica of what is obtainable in the wild (the fish's natural environment). However, this is not in common practice because not only are a lot of eggs lost as unfertilized in the process, also the timing cannot really be manipulated for results. The success rate of this method is low especially when considered from a commercial point of view.

The second method on the other hand, that is the **artificial method,** is more flexible and commercially viable. As a matter of fact, it is the method in common practice as the rate of success is higher compared to the natural pairing method. This involves inducing the female broodstocks with certain hormones towards getting their egg sacs loosed and ready for the breeding process as well as obtaining the milt needed for fertilization from the males.

7.3 Artificial (Hormone Induced) Breeding in Catfish

As earlier mentioned, artificial breeding involves inducing broodstocks (especially females) with certain chemicals called hormones; and this can be carried out in culturable fish species such as *Cyprinus carpio, Heterobranchus bidorsalis, Clarias gariepinus* etc. Although the process is very similar in these species, we shall use catfish as illustration. However, before breeding exercise takes place, certain facilities and materials need to be present. These include the hatchery structure, spawn tanks, broodstock holding tanks, saline solution etc.

7.3.1 The Hatchery Structure

The hatchery is the place where breeding takes place. It is usually indoors so as to protect the fries from predators and sudden weather changes. The fingerlings tanks are usually small and have efficient water inlet and outlet facilities. The hatchery could use static, flow through or recirculatory water system. The materials needed which are kept in the hatchery are: Broodstock holding tanks, broodstock holding bowls, broodstocks (male and female), hand towels, kitchen knife, scissors or blade, stripping bowls, 0.9% saline solution, hormone (pituitary or commercial types like ovaprim, ovulin, ovatide etc), needle and syringe, pH meter, thermometer, incubation troughs (which can be wooden vats, plastic tanks, etc), substrate (net tray, kakaban or basket), torchlight, live food (zooplankton) or processed food like shell free artemia, various grades of feed (0.2mm, 0.3 – 0.5mm, 0.5 – 0.8mm etc), well aerated water, tissue paper and of course standby personnel.

Hatchery operation is a delicate venture which must be handled carefully, hence mistakes must be avoided as much as possible. The hatchery building is usually built in such a way that the sides are covered with black tarpaulin or leather so as to maintain an optimum temperature in the hatchery.

7.3.2 The Breeding Process
This will be analyzed in steps for easy understanding.

<u>**Step 1:Broodstock Selection**</u>

Catfish gets sexually mature from six months. But parent stocks (broodstocks to be used for breeding) are usually set aside for at least a year to ensure proper development of the egg and milt.

Selection of female broodstocks for spawning starts with identifying a gravid fish, (gravid is like saying a "pregnant" fish). These are to be watched out for in selecting the gravid females:

1. Soft and protruded belly;
2. Release of brownish green eggs when the eggs are gently pressed out of the fish's belly.
3. Round shape of the genital opening.
4. Reddish colour of the fish's vent

These when observed point to the fact that the female broodstock is mature enough to be used. It is however worthy of note that though different species of catfish have their eggs in varying shades of colour green indicating maturity, the typical local catfish have deep brown eggs when mature as indicated in figure 7.1 below.

Fig. 7.1: Selection of a female Clarias
broodstock

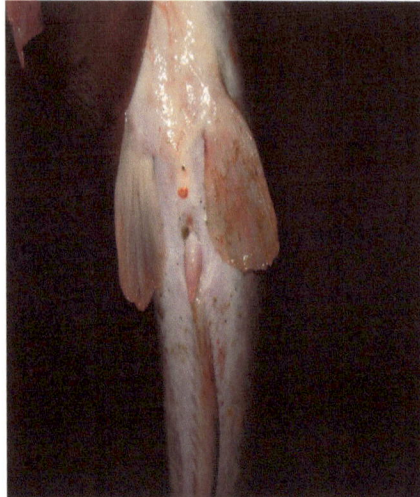

Fig. 7.2: Selection of a male Clarias broodstock

On the other hand, to select a mature male broodstock, the following are checked;

1. Longer or extended papillae (genital)
2. Reddish spot at the tip of the papillae
3. Slimmer outline unlike the females with protruded bellies
4. Milky coloured milt from the testes removed when cut open.

These when observed as displayed in figure 7.2 imply that the male should have good milt and therefore can be used for the breeding exercise. (Milky coloured milt is what usually indicates maturity in males).

After selecting the broodstocks based on these observations, they are to be kept in separate broodstock holding bowls so as to acclimatize them for the breeding purpose.

It is to be noted however that it is only good broodstocks that can produce good fish seeds; hence the source of procurement is to be carefully considered. Inquiries are to be made from technically sound service providers as well as successful hatchery operators so as not to procure bad broodstocks.

Step 2: Hormone Administration (Hypophysation)

Before this is properly explained, it is necessary to state that spawning is of two types in catfish, i.e. the pairing and the stripping methods. In the pairing method, both the male and female fish are injected with hormone and are paired up in a tank of water. After about 1 –12 hours the male fertilizes the eggs as the female lays them. This method is not in common practice because a lot of eggs are not fertilized and the success rate is low. As a result of this, consideration is given to the other method, which is in common practice and has good results. Here, the selected female broodstocks are weighed on a scale and then injected according to their weights. In practice, the hormone can be pituitary or the commercially available ones. For instance, in the case of ovaprim or ovulin hormone (which usually comes in a 10ml bottle), the weighed fishes are injected at a dosage of 0.5ml/kg of fish. As the weights vary, the dosages are adjusted as follows:

1kg of fish will take 0.5ml of the hormone
1.5kg of fish will take 0.75ml of the hormone
2kg of fish will take 1.0ml of the hormone
2.5kg of fish will take 1.25ml of the hormone and so on.

The injection is done intramuscularly just above the lateral line, and it can either be towards the head region or towards the tail region. However, both have been proven to be effective and as a result yield successful results.

The injection can be done early in the morning (6 – 7am) or late in the evening (9 – 10pm) so as to make room for the latency period (the waiting period from when the hormone is administered to when eggs are ready for stripping) which is usually 9 – 10 hours (although, this is dependent on the temperature of the water in which the fish is kept). For instance, at 28°c, it will take about 10 hours, while at 25°c it will take about 12 hours. The end of the latency period is however indicated by eggs flowing gradually out of the fish without any pressure. At this stage, the fish is gently removed from the water to prevent wasting of eggs.

It is to be ensured that during injection, a new syringe is used and the fish's skin is wet for proper penetration and prevention of injury on the fish. The injected point is also rubbed gently to prevent the back flow of the hormone as well as reducing the pain on the fish. The injected fish is then carefully returned to the holding bowl, which is covered with heavy materials (say like wood pallets, heavy planks etc) to keep it from jumping out and thus stressing or injuring itself during the latency period, as they are usually restless during this period.

Step 3: Stripping and Fertilization

After the latency period has elapsed, i.e. when the eggs have started running, the male fish is to be dissected or cut open (close to the heart) to get the testes.

The testes are carefully brought out and the milt they contain pressed into a clean bowl with 0.9% saline solution. The milt (sperm) should be white and milky in nature, as watery milt may not be viable for fertilization. A small bowl is then to be dried with a towel and the eggs pressed out (stripped) into it. This is done by gently pressing the belly of the fish in the tail direction to allow the eggs into the bowl through its vent. Two people are needed here, the person doing the stripping holds the head with a towel while the other holds the tail with a towel for firm grip as the eggs ooze out into the dry bowl.

The pressing is done until there are no more eggs coming out. After this, the milt, diluted in the saline solution to increase its surface area is then applied on the eggs by broadcast and mixed gently with a plastic spoon or feather for even fertilization. The fertilisation, usually described as wet fertilization, takes place within two (2) minutes of the milt's contact with the eggs. Little water is added to the medium and decanted to get out the surface foams and testes' sacs (if added), leaving the eggs clearly visible.

Step 4: Incubation of the Eggs

This is the process of consistently supplying oxygen to the fertilized eggs until they hatch. Oxygen supply can be by means of a mechanical air pump (aerator) or by allowing fresh water to flow through the tank. The fertilized eggs are spread on a single layer on the substrates which can be net trays or kakaban. These substrates serve as propagation beds for the fertilized eggs to become hatchlings i.e. they make allowance for easy separation of unhatched eggs from the hatchlings. The length of the incubation is a function of both the environmental and water temperature. Eggs hatch faster and better at higher temperatures, e.g. at 30°c, eggs will start hatching from the 16th hour while at 23°c, the eggs will start hatching at the 25th hour.

Step 5: Post Incubation Care

1. The temperature of the water is monitored with a thermometer so as to know when hatching is likely to commence. The aeration or water flow is increased immediately hatching is noticed so as to sustain the very tender hatchlings. The hatchlings will start dropping from the substrate into the water within

3 hours of the commencement of hatching. After about 24 - 28 hours from incubation time, when it is believed that all the hatchlings must have dropped off, the substrate is carefully removed into a basin containing water to prevent the unhatched eggs from also dropping into the culture medium. The removed substrate is left in the basin for about 3 hours to allow for late hatching during which the late hatchlings would swim up to the water surface and then taken back to the other ones. The aeration or water flow through continues to ensure clear water medium; cloudy water must be avoided by all means. The substrate can thereafter be washed, sun dried and properly kept for further use.

2. The fries normally feed on their yolk sacs (their natural internal food) till the third day after which they will be ready to feed on external feed – zooplankton. This could be processed artemia or naturally cultured ones like daphnia, moinea etc. The fries are moved out of where they hatched (incubation troughs) into other tanks or vats as the case may be to avoid ammonia interference and the likelihood of infection, which can lead to mortality. They are then well spaced out as well to encourage fast growth. As the fishes increase in size and age, they will be ready to take various grades of highly nutritional feeds to enhance their growth rate.

For instance, some of the commercially available Catfish Starter Feed come as indicated in Table 7.1 below - from hatchlings to 56 days (8 weeks) by average weight.

Table 7.1: Catfish Starter Feed Range by Average Body Weight

S/N	Age	Average Weight	Feed Type
1	0 – 7th day	0.025g	0.2mm or artemia

2	7 – 14th day	0.5 – 1.0g	0.2mm or artemia
3	14 – 21st day	1.0 – 1.5g	0.3 - .0.5mm
4	21 – 28th day	1.5 – 3.0g	0.3 – 0.5mm
5	28 – 35th day	3.0 – 5.0g	0.5 – 0.8mm
6	35 – 42nd day	5.0 – 9/10g	0.5 – 0.8mm
7	42 – 49th day	>10g	0.8 – 1.2mm
8	49 – 56th day	>12g	1.2 – 1.5mm

At this stage i.e. at 8 weeks old juveniles, they could be taken to the grow-out ponds where 2mm feed could be commenced for them. The feeding regime at the hatchery (with good water aeration) should be about 3 hours' interval.

3. Cleaning of the hatchery tanks. This is the removal of accumulated uneaten feed at the bottom of the tanks so as not to generate infections or ammonia build up and as such pollute the water, which can in effect lead to mortality (the fishes being very fragile at this stage). The exercise is generally referred to as siphoning, and it is usually done by sucking (bringing out) all the dirts in the tank by means of a pipette (hose) into a clean, salt free container. The dirts are to be discarded while the fishes that come out with them are cleaned and returned to the tanks. This exercise is faithfully and diligently carried out on a daily basis for the period of their stay in the hatchery, usually 3 weeks before being moved to the nursery tanks. During these periods in the hatchery, aeration must not stop. If it is a recirculatory system, the aeration must be continuous, and if it is flow through, aeration (through the inflow of fresh wa-

ter) must not cease as well. There must be consistent aeration throughout these first three weeks of their lives at the hatchery although they still need it as they grow on.

4. Removal of Shooters. During the course of their stay in the hatchery, the fries exhibit growth differences, which may either be due to genetic variation or feeding method. Some just emerge as shooters or jumpers, being bigger than the others. And to prevent cannibalism which is an intrinsic factor in catfish, a situation whereby the bigger ones feed on the smaller ones (being carnivorous fishes), the shooters are removed as soon as they are noticed and kept separately in other tanks or troughs, forming a population of their own. Shooters are not only to be removed once; they are to be removed as long as they are noticed until the others assume a uniform or an even growth. This issue of cannibalism should not be taken for granted as it can drastically trim down the fish population. Where there is no size difference, there will be no cannibalism.

5. Health Care. Proper hygiene is to be maintained in the hatchery. Just as nobody enters the theatre room or labour room and does anything he likes there, not everybody should be allowed into the hatchery room, let alone to do anything they like there. It is the "sacred place" of the whole fish production process; and neatness, orderliness, proper sanitation among other hygiene measures should be employed. The bowls and pipettes used are to be washed regularly with salt; the floor should always be kept clean and the personnel should also be conscious of proper hygiene in and out of the hatchery so as not to contaminate or bring infection into the system. In the case of the fishes, good water quality management will highly reduce the risk of infection or any disease condition. But in case of any infection or deviation from the proper health condition of the fishes or the culture medium, which can be indicated by smelly or cloudy water, hanging fishes, irregular

swimming, mortality etc medication can be given to arrest the situation. Such drugs that can be used include Aquaceryl plus, Fish vit plus, Fish biotics, Keproceryl WSP, $KMnO_4$ solution etc. They should be administered at their various prescribed dosages under an expert's supervision and monitored for effective performance. However, in case of any mortality, they should be quickly removed and disposed so as to prevent any chance of spreading infection. If good management can be ensured, the chances of having diseases will be very low.

7.4 Fish Breeding Calculations

The essence of this is to have an idea of what to expect at the end of a breeding exercise. Hatching percentage (hatchability) can only be evaluated by mere looking i.e. assumption based on observation. This is by comparing the amount of eggs hatched to those unhatched. It could be 50%, 60%, 70% etc but a higher percentage indicates a successful hatching.

It is worthy of note however that 10% of the female body weight is expected to be eggs; although it could sometimes be 15%.

Thus, for a fish of 1kg $\equiv$ 1,000g

10% will be 10/100 x 1000 = 100g

This means 100g of 1kg fish are eggs. But the number of eggs in 1g of fish is 500 – 700. Thus, using the least figure, we have

500 x 100g = 50,000 eggs (in 1kg fish),

With survival rate at 50%, we have

50/100 x 50,000 = 25,000 eggs

That is, 25,000 eggs are expected to survive out of the whole. It therefore follows that ideally, with these percentages, one should be able to realize 25,000 viable eggs from a fish of 1kg. So in knowing how many females to use in meeting a target production, we calculate with the 25,000 eggs/1kg fish.

For instance, to produce 2 million fingerlings or juveniles; we have;

2,000,000/25,000=80

This means 80 females of 1kg each would be needed to produce 2 million fingerlings or juveniles.

For the males, 2 can go for 5 females. So, if 'X' is the number of males needed,

2:5

X:80

5X = 80 x 2 (by cross multiplication)

5X = 160

X = 160/5

X = 32

That is, 32 males with good milt will be required for the production (with all other management factors in place). It is to be noted however that as the fishes grow, the population at hatchlings may tend to be reducing due to cannibalism or any other form of mortality. So, one doesn't always get the complete 25,000 fingerlings or juveniles from 1kg fish, it is just to guide the production towards meeting set targets. With this information in mind, farmers can effectively plan their breeding programme and evaluate their outcome at the end of each phase.

7.5 Required Levels of Important Water Parameters

To maintain good water quality for successful fish seeds production, certain parameters must be checked and worked with. They are:

1. **pH** - this should be between 6.5 and 8.0 (the best range for Clarias culture)
2. **Temperature** – this should be between 27°c – 30°c for good production
3. **Dissolved oxygen** – this should range between 5mg/l and 8mg/l
4. **Ammonia** – omg/litre is expected.
5. **Nitrite** – omg/litre expected.

These are just the important parameters, although there are more. A water quality test kit could be used to check these parameters or a sample of the water to be tested taken to a standard laboratory for analysis. It must be ensured that the ranges are safe before commencing the breeding exercise. Even upon starting i.e. when the production process is in progress, these parameters are to be checked regularly to ensure a stabilized culture medium.

With good broodstocks, good water quality, proper feeding at well-established regimes and effective hatchery management practices, a breeding exercise will be successful.

Study Questions:

1. Describe fish breeding.
2. Highlight five (5) processes involved in artificial fish breeding.
3. When are fishes said to be gravid?
4. What is latency period?
5. Differentiate between wet and dry fertilization.

Chapter Eight

ECONOMIC IMPORTANCE OF FISH AND FISH PRODUCTS

8.1 Fish and the Economy

Many industries in the world depend on the various raw materials from fisheries and aquaculture for finished products. Industries such as those involved in leather work also make use of fish skin and other water animals' skin for products like drums, shoes, bags, and wears materials. Some also process and can fish products. All these are put into household usage and market to earn domestically and internationally.

The fisheries and aquaculture sector of a nation also make use of some finished products from small and large scale industries like those into the manufacture of boats, ships, nets and other fishing equipment. These industries have greatly provided employment opportunities for those working there and have contributed immensely to the improvement of both domestic and foreign earnings, thereby improving the per capita income and standard of living in the country.

8.2 Fish and Science

Fish have been discovered to be of immense importance in the field of science. They are excellent research models in areas such as phylogenetics, evolutionary biogeography and ecology. The present fish fauna is living witness to climatic changes in the past, a fact that gives us information about past climate. Due to their life history traits, fishes have also been found to be suitable as early warning signals of anthropogenic stress on natural ecosystem dynamics, or conversely, as indicators of ecosystem recovery and of resilience. They are sensitive to many stresses from parasites to diseases to acidification. In this light, South American tropical fish *Apteronotus albifrons* (Gymnotiformes) have been proposed as biological early warning system to detect the presence of potassium cyanide in water by means of its electric organ discharges (EOD). With its neurogenic electric organ, this fish continually emits wave form electric signals, which are very stable under constant ambient conditions, but tend to vary in the presence of pollutants. This technique could be incorporated into a system for detecting changes in the quality of surface waters.

In biotechnology, fishes have been found to be useful as scientists have been able to extract antifreeze proteins from pond smelts (*Hypomesus nipponensis*) that can be used to protect the internal structure of products containing water (hydrated substances such as meat, vegetables, processed foods, blood, cells, tissues and organs).

Also, in the regulation of food web dynamics and nutrient balances, consumption of organisms by fish is a salient feature. This can regulate trophic structure and thus, influence the stability, resilience, and food web dynamics of aquatic ecosystems; changing as fish pass from one life stage to another. The feeding pattern of fishes can also influence the temporal availability of nutrients and the potential for algal blooms in nutrient-rich lakes, since fish mineralize nitrogen and phosphorus through excretion and defecation, thereby making these nutrients available for primary production.

In carbon flux regulation, fish communities can regulate the carbon-fixing capacity of nutrient-rich lakes, and thus, indirectly mediate the flux of carbon between a lake and the atmosphere. A study in Wisconsin, USA showed that a nutrient-rich lake with zooplanktivorous fish became a carbon sink, because zooplankton were suppressed and primary producers (carbon fixers) were released from grazing pressure.

Furthermore, fish communities, and specific species, are excellent indicators of biological and ecological integrity due to their continuous exposure to water conditions. They display an array of biotic responses, such as changes in growth, distribution and abundance related to water pollution, critical habitat degradation, eutrophication, organic enrichment, chemical toxicity, thermal changes and food availability and thus, should be key elements of ecosystem monitoring programs.

8.3 Other Importance of Fish and Fish Products

Apart from the aforementioned importance, fish and fish products have been discovered to be useful in several other ways. These include income generation, employment, human and animal nutrition, poverty alleviation and food security, contribution to GDP and balance of trade, disease management and control etc.

8.3.1 Human and Animal Nutrition

In a few countries in the world, fish consumption contributes up to 180 kcal per capita per day, but reaches such high levels only where there is lack of alternative protein foods grown locally or where there is a strong preference for fish, e.g Japan and some small island states. In industrialized countries, fish provides an average of 20-30 kcal per capita per day. Additionally, fish proteins are essential in the diet of some densely populated countries where the total protein intake level is low, and are very important in the diets of many other countries. The fact that fish is of high nutritional value is well known. Less well known however is the significant contribution that it makes to the diet of many fish-consuming communities in both the developed and developing world. Fish, especially marine fish, provides high quality protein and a wide variety of vitamins and minerals, including vitamins A and D, phosphorus, magnesium, selenium, and iodine. Fish is also a valuable source of essential fatty acids and its protein is easily digestible.

Even is small quantities, fish can have a significant positive impact on improving the quality of dietary protein intake by complementing the essential amino acids that are often present in low quantities in the rice and vegetable diets typical of many developing states. In particular, fish is a rich source of lysine which is an essential amino acid that is often deficient in rice diets with little animal protein.

Research has also shown that fish is more than just an alternative source of animal protein. Fish oils in fatty fish are the richest source of a type of fat that is vital for brain development in unborn babies and infants. This makes all fish and especially fatty fish, such as tuna, mackerel and sardine, particularly good components of the diet of pregnant and lactating women.

In addition to these, certain components such as fish meal and bone meal are ingredients used in compounding feeds for animals such as birds, pigs, fish etc. These ingredients serve as sources of animal protein and minerals in the feeds of such animals.

8.3.2 Income Generation

It is a well-known fact that most individuals become fishers or fish farmers because they expect the activity to provide them a means of livelihood as well as their families. Engaging in commercial fish farming serves as source of income as the farmer may culture fingerlings or juveniles, broodstocks, table size fish etc for sale. As these are sold, the farmer receives income. Also, for those engaged in capture fisheries, they go for a catch in the wild; and upon catching, they sell, thereby receiving income as well.

8.3.3 Employment

Fish farming has in recent times become a very important source of employment. Many graduates from institutions of learning now engage in this venture, and as a result are no longer part of the teeming population of the unemployed. Aside the fact that many now engage in fish production as a vocation, the activities involved in the venture also create opportunities for others (like farm managers, farm attendants, store keepers etc) to be employed.

In addition to all these, fish may also be collected live for research, observation, or for aquarium trade.

8.4 Some Fish Products and their Uses

Several fish products exist, many of which have been discovered to be useful in several ways ranging from health, food, aesthetics to industrial uses.

Table 8.1 gives a list of some of these fish products and their respective uses.

Table 8.1: Fish Products and their Uses

S/N	FISH PRODUCT	USES
1.	Fish oil	Used for medicinal purposes and also industrial purposes like in the production of soap, candle, lubricants etc
2.	Fish bones	Used for making fertilizers and also in compounding animal feeds as source of minerals.
3.	Fish meal	Used in compounding livestock feeds as source of animal protein.
4.	Fish scales and teeth	Used for ornamental purposes – as articles for decoration.
5.	Fish skin	Used for the production of leather belts, shoes, wallets and bags. Also they are used as abrasive.

6.	Fish shells	Used to reinforce building walls especially when those of oysters and periwinkle are mixed with cement and sand.
7.	Fish bladder	Used in producing wine.
8.	Omega -3 fatty acids	Used medically in addressing cardiovascular diseases.
9.	Fish glue	Obtained by boiling the skin, bones and swim bladders of fish. Used as gumming substance.
10.	Fish emulsion	Used mainly for industrial purposes.

Study Questions:

1. Briefly describe the importance of fish in the field of science.
2. Highlight six (6) fish products and their respective uses.
3. In what ways can fish serve as source of income and employment?

Glossary

Alkalinity – this is the concentration of basic minerals in the pond.

Amino acids – these are the biochemical "building blocks" of protein.

Aquaculture – this is the rearing or culture of aquatic organisms (flora and fauna) in a controlled water environment.

Benthic – fishes are said to be benthic when they mostly dwell at the bottom part of the water body.

Biodiversity – this refers to the variety of living things (organisms) in an environment.

Bony fish - these are fish with bony skeletal framework, cycloid scales and apical mouth part.

Breeding - fish breeding simply refers to the propagation of fish seeds by pairing or crossing sexually mature male fish with sexually mature female fish. It involves engaging the gametes of sexually mature male and female fish towards seeds production.

Broodstock – these are sexually mature male and female fish usually engaged in the breeding process.

Cannibalism – this is a condition whereby a particular fish (usually bigger) preys on others (usually smaller). It is an inherent factor in carnivorous species and usually encouraged by size differences.

Carnivorous – fishes are said to be carnivorous when they feed on flesh (animals).

Cartilaginous fish – these are fish with paired fins, placoid scales, two chambered hearts, ventral mouth and skeleton made up of cartilage.

Dissolved oxygen – this is the amount of atmospheric oxygen dissolved in water for easy absorption of aquatic organisms. It is so essential to life.

FCR – this means *Feed Conversion Ratio*. It is the amount of feed needed to produce one kilogram of flesh in fish. It is a measure of how efficient a feed is.

Fertilization – this is the process of introducing the milt of the male fish to the eggs stripped from (or laid by) the female.

Filleting – this is the process of removing bones from a fish.

Fisheries – these simply refer to all processes involved in fish production, processing, marketing and distribution.

Fishing craft – these are equipment which provide platform for fishing operations, carrying the crew and fishing gears.

Fishing gear – these are equipment (like nets) used for catching fish.

Fish oil – the oil extracted from the liver of some fish species or obtained during fish processing.

Fry – this refers to a hatchling fish (pre-fingerling size).

Gravid – female fishes are said to be gravid when they are carrying ripe eggs in their abdomen.

Habitat – this refers to the natural environment of organisms.

Hardiness – this is the ability of fish to withstand handling conditions and be adaptive to the associated stress.

Hardness – this refers to the ability of water to precipitate soap.

Herbivorous - fishes are said to be herbivorous when they feed on plant materials.

Host – an organism on or in which a parasite lives.

Incubation - this is the process of consistently supplying oxygen to the fertilized eggs spread on a substrate until they hatch.

Infectious disease – this is a disease resulting from the invasion of a pathogenic organism.

Intramuscular injection – this refers to the kind of injection administered within a muscle.

Latency period – this is a period female broodstocks pass through between the moment they are artificially induced (injected) and the time their eggs become ready for stripping. That is, the period between hormone administration and readiness for stripping.

Milt – this is the milky fluid obtained from the testes of mature male fish, used in fertilizing ripe fish eggs. It is the "sperm" of the fish.

Mortality – this is the term used to describe the death of living organisms.

Neoplastic disease – this refers to the abnormal growth or tumor that may appear in any part of the body with loss of structural and functional ability of the affected organ.

Non infectious disease – this refers to a disease or sickness not related to the activities of a virus, bacteria or other microorganism.

Nutritional disease – these are diseases associated with dietary deficiencies (unbalanced diet), improper processing and storage of feed and feed ingredients as well as feed contamination (chemical or biological).

Omnivorous – fishes are said to be omnivorous when they feed on both plants and animals.

Pathogens – simply put, these are disease causing organisms.

Pelagic – fishes are said to be pelagic when they mainly utilize the surface part of the water body.

pH - this is a measure of the alkaline and acid content of water given by the level of hydrogen ion (H+) dissolved in the water. pH means *potential of Hydrogen*.

Phytoplankton – these are plant micro organisms. That is, tiny plant life forms.

Polyculture – this is the practice of rearing or culturing two or more different species of fish together in the same water environment.

Pollution – this is the process of introducing dirty, harmful or poisonous substances into the water body thereby rendering it inconducive and unhealthy for the life forms living in it.

Seechi disc – this is an instrument used in measuring water turbidity.

Shelf life – this refers to the period a processed food can stay before spoilage.

Stripping – this is the process of extracting or collecting ripe eggs from the female fish by gently pressing its abdomen.

Stocking – this simply refers to the introduction of fish stock into a culture facility (fish pond or tank).

Taxonomy - Taxonomy is the practice and science of classification. It is composed of two Greek words, *taxis* (order, arrangement) and *nomos* (law or science); hence the word literally means "Science of arrangement.

Temperature – this refers to the degree of hotness or coldness of a water medium.

Turbidity – this is the measure of suspended matter in the pond water. It is denoted by absence of transparency in the water.

Zooplankton – these are animal micro organisms. That is, tiny animal life forms.

Reference and Further Reading

Brigitte, M., Brigiet, B. and Corlien, H. (1994). Preservation of Fish and Meat, Netherlands: AGROMISA Publishers.

David R Blakely and Christopher T. Hrusa (Inland Aquaculture Development Handbook)

Jackson, Jeremy B C et al. (2001) *Historical overfishing and the recent collapse of coastal ecosystems* Science 293:629-638.

Jedida, A.N. (2011). Manual for African Catfish (*Clarias gariepinus*) Breeding in Zambia (online document).

Kwarteng, J.A and Towler, M.J. (1994). Western African Agriculture, a Textbook for Schools and Colleges, London: The Macmillan Press Ltd.

Practical Manual for the Culture of the African Catfish (*Clarias gariepinus*).

Joint publication of Directorate General International Cooperation of the Ministry of Foreign Affairs, The Hague, the Netherlands; Department of Fish Culture and Fisheries of the Agricultural University of Wageningen, the Netherlands and Research Group for Comparative Endocrinology, Department of Zoology of the University of Utrecht, the Netherlands.

About the Author

Prince Anthony Adefarakan as he is popularly called is the M.D/ CEO of Aquaton Konsults, Nigeria, West Africa. He is an experienced Fisheries Consultant with vast wealth of knowledge in matters relating to fish production. He has practically demonstrated artificial fish breeding, fish ponds construction, fish feeding, fish disease and management among other fish production techniques to a large number of farmers far and wide. In addition to training these farmers (some of which are students, retirees and investors), he has been personally involved in their business set up; providing the necessary resources to ensure their success.

At some point in his Fisheries career, he served as a Master Aquaculture Service Provider (MASP) to a Department for International Development (DFID) funded project in the Niger Delta part of Nigeria (Market Development Project in the Niger Delta). He also served as one of the USAID's Nigerian Agricultural Enterprise Curriculum (NAEC) trainers in the Niger Delta.

His impact was felt in academics as well. He was a lecturer and also the personnel appointed to handle the Fisheries section of the World Bank funded STEP-B Project of the Federal College of Education (Technical), Asaba where he had the opportunity to impact the Agriculture students of the institution with relevant aquaculture knowledge capable of making them self-reliant upon graduation. Furthermore, he has had the opportunity to serve as one of the Fisheries Examiners and Moderators for West African Examination Council.

This book is therefore a compressed presentation of both his theoretical and field expertise. For successful fish production at all levels, this handbook is a must read
He currently lives in Canada with his wife and son.

www.ingramcontent.com/pod-product-compliance
Lightning Source LLC
Chambersburg PA
CBHW041159220326

41597CB00001BA/13